MAPPING WORKBOOK

Karl Byrand

University of Wisconsin Colleges

GLOBALIZATION AND DIVERSITY
GEOGRAPHY OF A CHANGING WORLD

FOURTH EDITION
Rowntree • Lewis • Price • Wyckoff

Boston   Columbus   Indianapolis   New York   San Francisco   Upper Saddle River
Amsterdam   Cape Town   Dubai   London   Madrid   Milan   Munich   Paris   Montréal   Toronto
Delhi   Mexico City   São Paulo   Sydney   Hong Kong   Seoul   Singapore   Taipei   Tokyo

Senior Geography Editor: Christian Botting
Senior Marketing Manager: Maureen McLaughlin
Assistant Editor: Kristen Sanchez
Editorial Assistant: Bethany Sexton
Managing Editor, Geosciences and Chemistry: Gina M. Cheselka
Project Manager: Pat Brown
Composition: PreMediaGlobal
Cover Image: Adapted from DK World Atlas, London: DK Publishing, 1997.

www.pearsonhighered.com

1 2 3 4 5 6 7 8 9 10—EBM—17 16 15 14 13
ISBN-10: 0-32-186220-1; ISBN-13: 978-0-32-186220-4

Table of Contents

Chapter One: Concepts of World Geography Mapping Workbook Exercises

Identify the following world regions on Mapping Workbook Map 1.1

Australia and Oceania
Central Asia
East Asia
Europe
North Africa/Southwest Asia
North America
Latin America
South Asia
Southeast Asia
Sub-Saharan Africa
The Caribbean
The Russian Domain

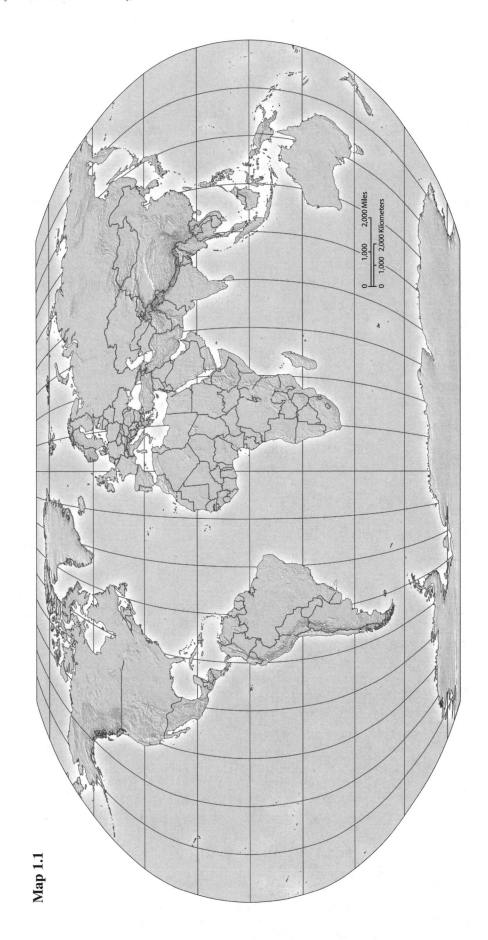

Map 1.1

Exercise One: Population Indicator Comparison

Using Table 1.1 "Population Indicators" (p. 23) and Mapping Workbook Map 1.2, complete the following exercise.

Using the data provided in the Population Density (per square kilometer density) column of Table 1.1 "Population Indicators" (p. 23), use a colored pencil to shade in the countries with a per kilometer density of 8.0–77.6. Using different colored pencils for each category, shade the countries that possess a per kilometer population density of 77.7–147.2, 147.3–216.8, 216.9–286.4, and 286.5 and above. Mark your shading scheme in the map's legend. Once you have done this, answer the following questions.

1. Of the mapped countries, which have the highest population density?

_____

2. Is there a regional pattern associated with countries possessing high population densities?

_____

3. Of the mapped countries, which have the lowest population density?

_____

4. Does there appear to be a regional pattern associated with countries possessing low population densities?

_____

Compare those countries you shaded with the population distributions displayed in Figure 1.24 "World Population" (p. 20).

5. Would you say that the population is evenly distributed (i.e., dispersed) within the majority of countries, or does it appear to display a concentration?

_____

6. Explain the reasons for either observed pattern. More specifically, what common landscape characteristics (e.g., consider the physical geography of the regions where there are high and low population densities) tend to influence where population is concentrated?

_____

_____

_____

_____

_____

_____

Using the data provided in the total fertility rate (TFR) column of Table 1.1 "Population Indicators" (p. 23), use a colored pencil to pattern in the countries with a TFR of 1.4–2.26. Using different colored pencils for each category, pattern the countries that possess TFRs of 2.27–3.12, 3.13–3.98, 3.99–4.84, 4.85–5.7. Mark your patterning scheme in the map's legend. Once you have done this, answer the following questions.

7. Compare your population density map with the map of TFRs you created. Does there appear to be a correlation between countries possessing high TFRs and high population densities? Please specify which countries display this pattern.

_____

_____

_____

_____

_____

8. Does there appear to be a correlation between countries possessing low TFRs and low per square kilometer population densities? Please specify which countries display this pattern.

_____

_____

_____

9. Are there outliers to either of these scenarios (i.e., countries with high population densities and low TFRs, and vice versa)? If so, name these countries and explain the possible reasons why this is the situation.

_____

_____

_____

_____

_____

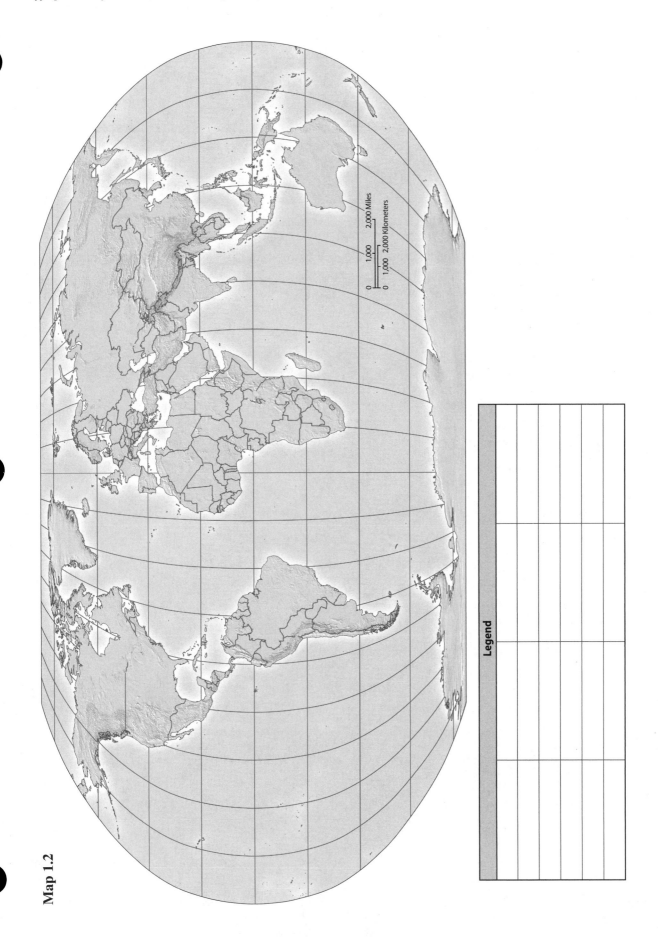

Map 1.2

Legend

Exercise Two: Development Indicator Comparison

Using Table 1.2 "Development Indicators" (p. 37) and Mapping Workbook Map 1.3, complete the following exercise.

Using the data provided in the Gross National Income (GNI) Per Capita, Purchasing Power Parity (PPP) 2010 column of Table 1.2 "Development Indicators" (p. 37), use a colored pencil to shade in the countries with a GNI of $1,400–$2,400. Using different colored pencils for each category, shade the countries that possess GNIs of $2,401–$4,400, $4,401–$6,400, $6,401–$10,400, and $10,401 and above. Mark your shading scheme in the map's legend. Once you have done this, answer the following questions.

Using the data provided in the Human Development Index (HDI), 2011, column of Table 1.2 "Development Indicators" (p. 37), use a colored pencil to shade in the countries with a HDI of 0.511–0.6008. Using different colored pencils for each category, shade the countries that possess HDIs of 0.6009–0.6906, 0.6907–0.7804, 0.7805–0.8702, and 0.8703–0.960. Mark your shading scheme in the map's legend. Once you have done this, answer the following questions.

1. Define GNI.

    _____

    _____

2. Define HDI.

    _____

    _____

3. Among the 10 most populous countries, is there a regional pattern of high, medium, and low GNI countries? If so, what is that pattern?

    _____

    _____

    _____

4. Among the 10 most populous countries, is there a regional pattern of high, medium, and low HDI countries? If so, what is that pattern?

    _____

    _____

    _____

5. Now compare the maps. What correlation is there between countries whose development is measured by GNI as compared with HDI? For example, do the same countries that possess a high HDI also display a high GNI and countries that display a low HDI display a low GNI?

    _____

    _____

    _____

6. Name the countries that display this correlation.

_____

_____

_____

7. Comparing the maps once again, do you notice outliers to what you observed in question 5? Specifically, are there countries that have a different ranking with their GNIs and their HDIs?

_____

_____

_____

8. Name any countries that display this pattern.

_____

_____

_____

9. What do you think would account for such differences in GNI and HDI in outlier countries?

_____

_____

_____

10. Based on this, which do you believe is a better measure of development? Why?

_____

_____

_____

_____

_____

_____

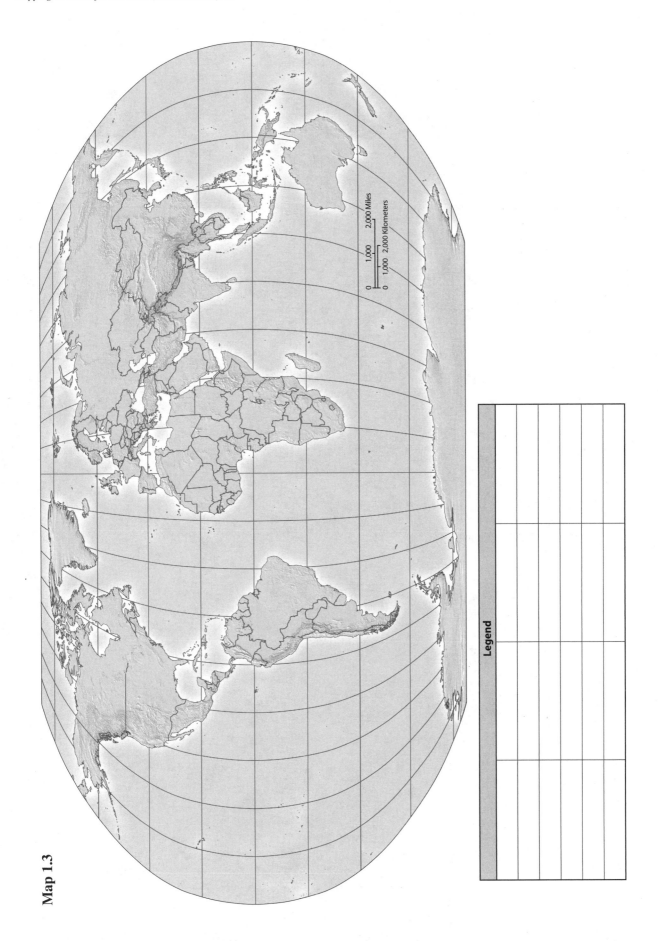

Map 1.3

Legend

Bonus Exercise: Create a third map based on the development indicator of your choosing from Table 1.2 and compare it against the GNI and HDI of the 10 most populous countries. Discuss the correlations, the outliers, and the reasons why they occur.

Table

GNI	Color	HDI	Color
$1,400–$2,400		511–0.6008	
$2,401–$4,400		0.6009–0.6906	
$4,401–$6,400		0.6907–0.7804	
$6,401–$10,400		0.7805–0.8702	
$10,401 and above		0.8703–0.960	

Chapter Two: The Changing Global Environment Mapping Workbook Exercises

Exercise One: Regional Impacts of Global Warming

Using Figure 2.21 "Largest CO_2 Emitters" (p. 55) and Mapping Workbook Maps 2.1, 2.2, and 2.3, complete the following exercise.

Using the data provided in Figure 2.21, the "Largest CO_2 Emitters" graph (p. 55), use a colored pencil to shade in the countries with a 2011 CO_2 emission of 0.5–1.0 gigatons per year. Using different colored pencils for each category, shade the countries that possess a 2011 CO_2 emission of 1.1–2.0 gigatons per year, and a 2011 CO_2 emission of 2.1 or greater gigatons per year. Mark your shading scheme in the map's legend. Once you have done this, answer the following questions.

1. Is there a geographic pattern associated with these polluting countries (e.g., a regional concentration, or a more random distribution)? If so, what is it?

   _____

   _____

2. Do these nations possess the same level of economic and industrial development? Elaborate.

   _____

   _____

Now using the data provided in Figure 2.21, the "Largest CO_2 Emitters" graph and maps 2.2 and 2.3, do the same as above for the 2000 and 1990 CO_2 emissions data. As you produce your two other maps, use the same data categories and shading scheme.

3. As compared with the map you generated using the 2011 data, does the geographic pattern associated with these polluting countries differ in 2000? If so how does it differ?

   _____

   _____

4. Does the pattern differ between 1990 and 2000? If so, how does it differ?

   _____

   _____

Now read the section "Deforestation of Tropical Forests" (p. 58) and study Figure 2.25 "World Bioregions" (pp. 58–59) and note the locations of Earth's tropical rain forests.

On your 2011 map (Mapping Workbook Map 2.1) use a red or black (or another color that will contrast with the shading scheme you have created) colored pencil to pattern in the countries/regions experiencing the highest levels of tropical deforestation. Mark this shading scheme in the map's legend.

Now read the section "Deserts and Grasslands" (p. 60) and study Figure 2.25 "World Bioregions" (pp. 58–59) and note the locations of Earth's deserts and grasslands.

On your 2011 map (Mapping Workbook Map 2.1) use a red or black (or another color that will contrast with the shading and patterning schemes you have created) colored pencil to pattern in the countries/regions experiencing the highest levels of desertification. Mark your shading scheme in the map's legend. Once you have done this, answer the following questions.

5. What are the primary causes of tropical deforestation?

_____

_____

6. How does tropical deforestation contribute to and exacerbate global atmospheric change (i.e., global warming)?

_____

_____

7. What are the primary causes of desertification?

_____

_____

8. How does desertification contribute to and exacerbate global atmospheric change (i.e., global warming)?

_____

_____

Compare the areas that are experiencing tropical deforestation and desertification with the pattern of the major CO_2 producers.

9. Is there a correlation between these two patterns (e.g., are these producers the same nations that are experiencing the highest rates of tropical deforestation and desertification)?

_____

_____

Another significant problem associated with global warming is sea level rise. On your map, use a red or black (or another color that will contrast with the shading and patterning schemes you have established) colored pencil to pattern the coastal areas of those countries that are the major CO_2 producers. Mark your shading scheme in the map's legend.

Now examine Figure 1.24 "World Population" (p. 20). Once you have done this, answer the following questions.

10. Is there a correlation between the regions of highest population density and coastal areas in the major CO_2-producing countries? If so, what is it?

_____

_____

11. It is estimated that sea levels could rise 4 feet (1.2 meters) by the year 2100. If this holds true, how will sea level rise have an impact on these CO_2-producing nations?

_____

_____

12. Will there also be an impact on the nations that produce only small quantities of CO_2?

_____

_____

13. How equitable is this situation?

_____

_____

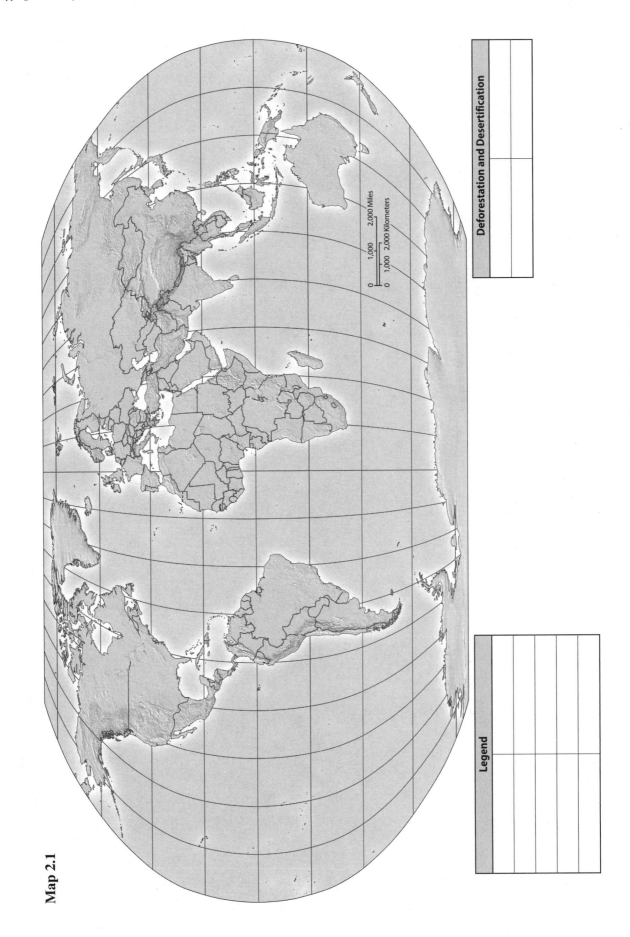

Map 2.1

Deforestation and Desertification

Legend

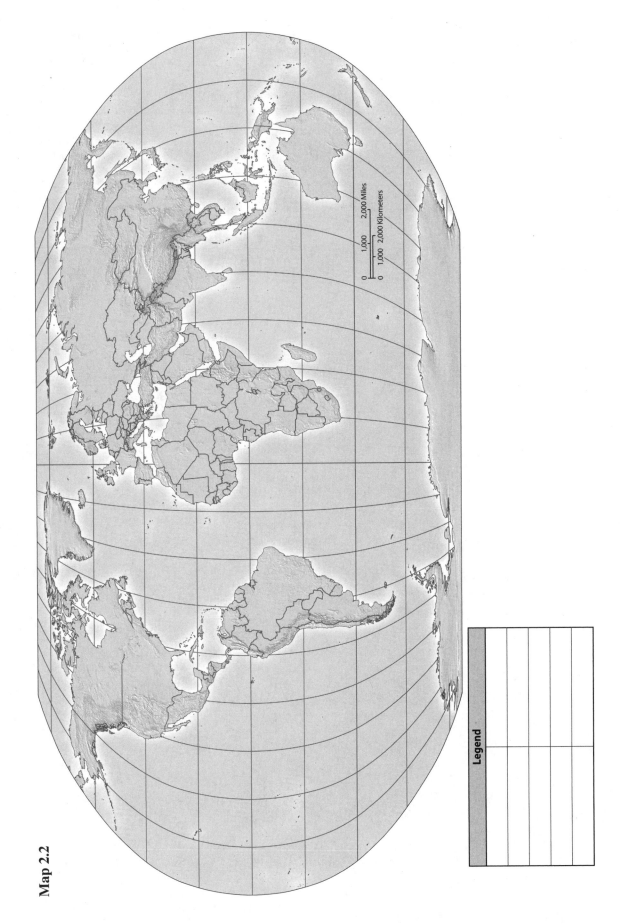

Map 2.2

2,000 Miles

1,000   2,000 Kilometers

0

Legend

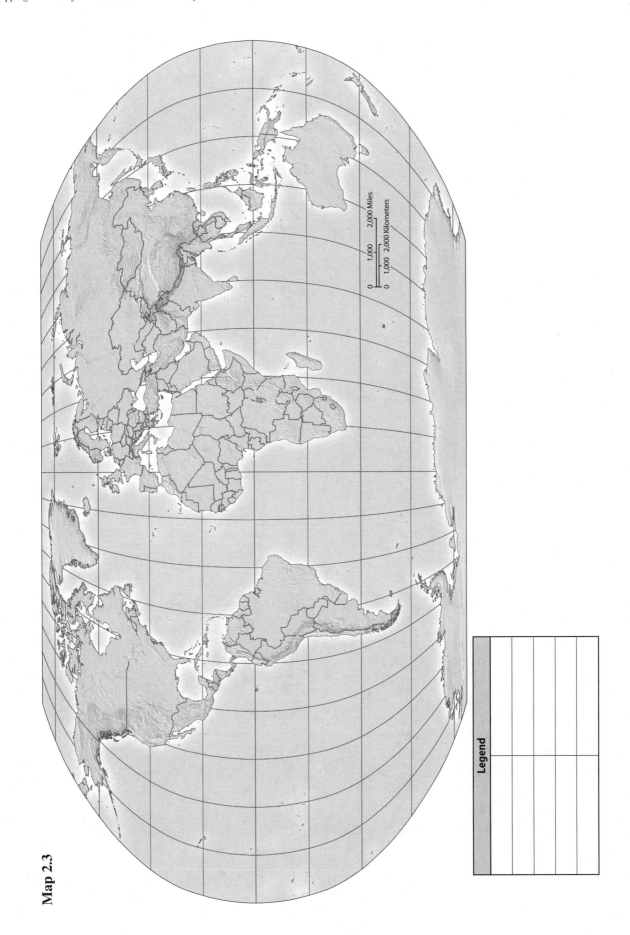

Map 2.3

Legend

Exercise Two: Climate as a Biome Control

Using Figure 2.18 "World Climate Regions" (pp. 52–53), Figure 2.25 "World Bioregions" (pp. 58–59), Figure 1.24 "World Population" (p. 20), and Mapping Workbook Map 2.4, complete the following exercise.

Using Figure 2.18 "World Climate Regions" (pp. 52–53) as a reference, use different colored pencils to shade in the approximate locations of the major climatic regions. Mark your shading pattern in the map's legend. Using Figure 2.25 "World Bioregions" (pp. 58–59) as a reference, using different colors to represent each bioregion, shade in the various bioregions in their respective locations. Mark your patterning scheme in the map's legend. Once you have done this, answer the following questions.

In the table below, specifically match each bioregion with its mapped climatic region.

Bioregion	Climate region

1. What is the general pattern of world climatic regions?

_____

_____

_____

2. What are the general controls (i.e., causes or determinants) of these climate regions?

_____

_____

_____

3. What is the general correlation between the climatic regions and the world bioregions?

_____

_____

_____

4. What characteristics of these climatic regions determine these bioregions?

_____

_____

_____

Using Figure 1.24 "World Population" (p. 20) as a reference, use a red or black (or another color that will contrast with the shading and patterning schemes you have established) colored pencil to shade in the regions possessing the highest population density.

5. In which climate region(s) do(es) the majority of the planet's population reside?

_____

_____

_____

6. What are the characteristics of this (these) climatic region(s)?

_____

_____

_____

7. Why do you think the majority of the planet's population resides within this (these) regions(s)?

_____

_____

_____

Examine the regions of low population density depicted in Figure 1.24 "World Population" (p. 20).

8. In which climate region(s) do we see these lower population densities?

_____

_____

_____

9. What are the characteristics of this (these) climatic region(s)?

_____

_____

_____

10. Why are there lower population densities in these climatic regions?

_____

_____

_____

Map 2.4

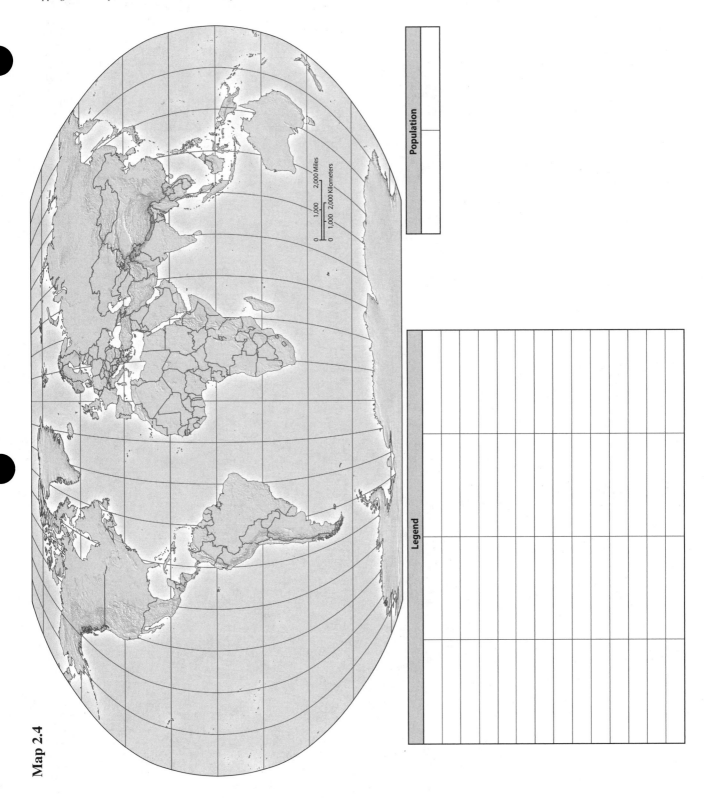

Population

Legend

Chapter Three: North America Mapping Workbook Exercises

Identify the following features on Mapping Workbook Maps 3.1, 3.2, and 3.3

Identify and label the following countries on Mapping Workbook Map 3.1

Canada
United States

Identify and label the following states and provinces on Mapping Workbook Map 3.1

Alabama	Maryland	Ontario
Alaska	Massachusetts	Oregon
Alberta	Michigan	Pennsylvania
Arizona	Minnesota	Prince Edward Island (P.E.I.)
Arkansas	Mississippi	Quebec
British Columbia	Missouri	Rhode Island
California	Montana	Saskatchewan
Colorado	Nebraska	South Carolina
Connecticut	Nevada	South Dakota
Delaware	New Brunswick	Tennessee
Florida	New Hampshire	Texas
Georgia	New Jersey	Utah
Hawaii	New Mexico	Vermont
Idaho	New York	Virginia
Illinois	Newfoundland and Labrador	Washington
Indiana	North Carolina	Washington, DC
Iowa	North Dakota	West Virginia
Kansas	Northwest Territories	Wisconsin
Kentucky	Nova Scotia	Wyoming
Louisiana	Nunavut	Yukon Territory
Maine	Ohio	
Manitoba	Oklahoma	

Identify and label the following cities on Mapping Workbook Map 3.2

Albuquerque	Kansas City	Phoenix
Anchorage	Las Vegas	Portland
Atlanta	Los Angeles	Quebec City
Boston	Louisville	Regina
Calgary	Memphis	Salt Lake City
Charlotte	Miami	San Diego
Chicago	Milwaukee	San Francisco
Columbus	Minneapolis–St. Paul	Seattle
Dallas–Fort Worth	Montreal	Saint John
Denver	Nashville	St. Louis
Detroit	New Orleans	Tampa
Halifax	New York City	Toronto
Honolulu	Oklahoma City	Whitehorse
Houston	Ottawa	Winnipeg
Indianapolis	Philadelphia	Vancouver

Identify and label the following physical features on Mapping Workbook Map 3.3

Alaska Range
Appalachian Highlands
Arctic Ocean
Atlantic Ocean
Baffin Bay
Baffin Island
Bering Strait
Brooks Range
Cascade Range
Central Valley
Chesapeake Bay
Coast Mountains
Coast Ranges
Colorado Plateau
Columbia River
Florida Keys

Great Plains
Gulf of Alaska
Gulf of Mexico
Great Basin
Hawaii
Hudson Bay
Kauai
Lake Erie
Lake Huron
Lake Michigan
Lake Ontario
Lake Superior
Mackenzie River
Maui
Mississippi River
Missouri River

Newfoundland
Oahu
Ohio River
Ouachita Plateau
Ozark Mountains
Pacific Ocean
Piedmont
Prudhoe Bay
Rio Grande River
Rocky Mountains
Sierra Nevada
St. Lawrence River
Teton Range
Vancouver Island

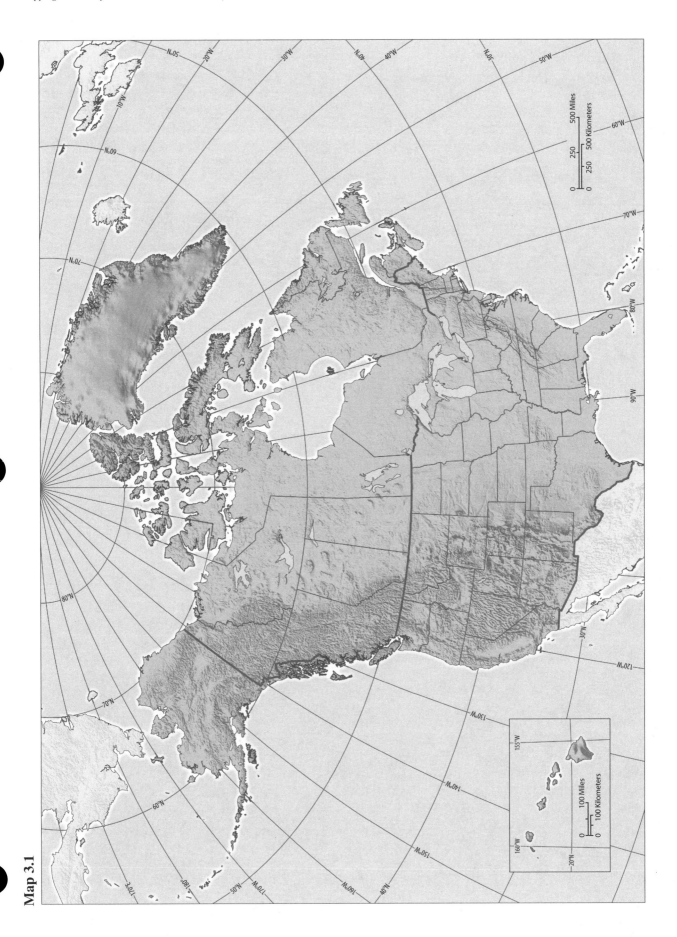

Map 3.1

Map 3.2

Map 3.3

Exercise One: Internal Patterns of North American Migration

Using Figure 3.1 "North America" (p. 69), Figure 3.14 "European Settlement Expansion" (p. 80), and Mapping Workbook Map 3.4, complete the following exercise.

Read the section "Occupying the Land" (pp. 79–80) in your textbook. Using Figure 3.14 "European Settlement Expansion" (p. 80) as a reference, create an isochronic (line of equal time) map. Begin by using a colored pencil and draw a line indicating the North American regions that were settled by 1750. Label this line "1750." Using a different colored pencil, draw a line indicating the North American regions that were settled between 1750 and 1850. Label this line "1750–1850." Using a different colored pencil, draw a line indicating the North American regions that were settled by 1910. Label this line "1910." Mark the lines in the map's legend. Once you have done this, answer the following questions.

1. Prior to 1750, what were the major North American regions that were settled?

_____

2. What are the factors explaining this settlement pattern?

_____

_____

3. What were the major North American regions settled between 1750 and 1850?

_____

_____

4. What are the factors explaining this settlement pattern?

_____

_____

5. What were the major North American regions settled between 1850 and 1910?

_____

_____

6. What are the factors explaining this settlement pattern?

_____

_____

Examine Figure 3.1 "North America" (p. 69), paying close attention to the continent's physical geography.

7. How do you think North America's physical landscape has had an impact on the 1750, 1750–1850, and 1910 settlement patterns?

_____

_____

8. At what point in time, if any, do you believe that the physical landscape would have been less of a determinant associated with this pattern of westward settlement?

_____

_____

9. If you believe that the physical landscape eventually became less of a determinant, explain the reasons behind its decreased impact.

_____

_____

Read the section "North Americans on the Move" (pp. 80–82) in your text. Using a colored pencil, create flow arrows indicating the movement of Blacks from the rural South to the urban North. Label the arrow with the approximate period that this migration occurred. Mark this symbol in the map's legend.

Using a different colored pencil, create a flow arrow indicating the general pattern of regional growth in the Sun Belt South. Label the arrow with the approximate period that this migration occurred. Mark this symbol in the map's legend.

Using a different colored pencil, create a flow arrow indicating the general pattern of regional growth in the mountain states. Label the arrow with the approximate period that this migration occurred. Mark this symbol in the map's legend. Once you have done this, answer the following questions.

10. What was the major period when Blacks migrated from the rural South to the urban North?

_____

11. What were the contributing factors associated with this migration?

_____

_____

12. What were the impacts of this migration?

_____

_____

13. What was the major period when the Sun Belt witnessed the largest growth?

_____

14. What were the contributing factors associated with this migration?

_____

_____

15. What were the impacts of this migration?

_____

_____

16. What was the major period when the mountain states witnessed growth?

_____

17. What were the contributing factors associated with this migration?

_____

_____

18. What were the impacts of this migration?

_____

_____

Exercise Two: United States Patterns of Hispanic and Asian Residence

Using Figure 3.22 "Distribution of U.S. Hispanic and Asian Populations, by State, 2010" (p. 85) and Mapping Workbook Maps 3.5 and 3.6, complete the following exercise.

Using Figure 3.22 "Distribution of U.S. Hispanic and Asian Populations, by State, 2010" (p. 85) as a reference, use a colored pencil to shade in the states that possess greater than 30 percent of the Hispanic households in the United States. Using a different colored pencil, shade in the states that possess between 18 percent and 20 percent of the Hispanic households in the United States. Using a different colored pencil, shade in the states that possess between 7 percent and 9 percent of the Hispanic households in the United States. Using a different colored pencil, shade in the states that possess between 2 percent and 5 percent of the Hispanic households in the United States. Mark your shading scheme in the map's legend.

Examine the "Asian Household by Population, by State" pie chart and use a colored pencil to pattern in the states that possess greater than 30 percent of the Asian households in the United States. Using a different colored pencil, pattern in the states that possess between 18 percent and 25 percent of the Asian households in the United States. Using a different colored pencil, pattern in the states that possess between 7 percent and 10 percent of the Asian households in the United States. Using a different colored pencil, pattern in the states that possess between 2 percent and 5 percent of the Asian households in the United States. Mark this patterning scheme in the map's legend. Once you have done this, answer the following questions.

1. Which U.S. states possess the highest percentages of Hispanics?

_____

_____

2. What might account for the higher percentages in these states?

_____

_____

3. Which U.S. states possess the lowest percentages of Hispanics?

_____

_____

4. What do you believe accounts for the lower percentages of Hispanics in these states?

_____

_____

5. Which U.S. states possess the highest percentages of Asians?

_____

_____

6. What do you believe accounts for the higher percentages in these states?

_____

_____

7. Which U.S. states possess the lowest percentages of Asians?

_____

_____

8. What do you think accounts for the lower percentages of Asians in these states?

_____

_____

9. Which states possess a correlation between larger percentages of Hispanics and larger percentages of Asians?

_____

_____

10. What do you think are the factors responsible for this correlation?

_____

_____

11. Which states possess *higher* percentages of Hispanics and *lower* percentages of Asians?

_____

_____

12. What might be the contributing factors associated with this reverse correlation? If so, what might those factors be?

_____

_____

_____

_____

_____

13. Which states possess *lower* percentages of Hispanics and *higher* percentages of Asians?

_____

_____

14. Do you think that there are any contributing factors associated with this reverse correlation? If so, what might those factors be?

_____

_____

_____

_____

_____

_____

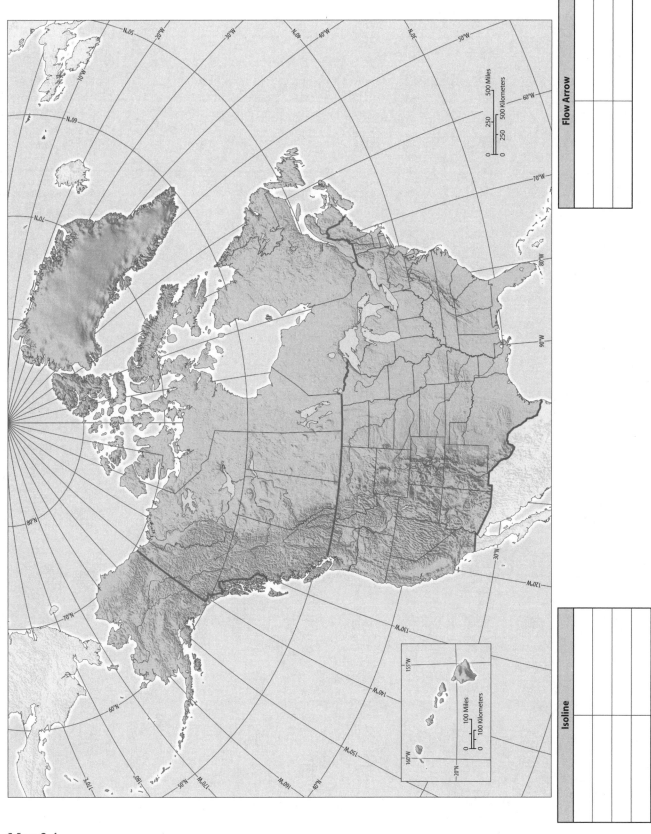

Flow Arrow

Isoline

Map 3.4

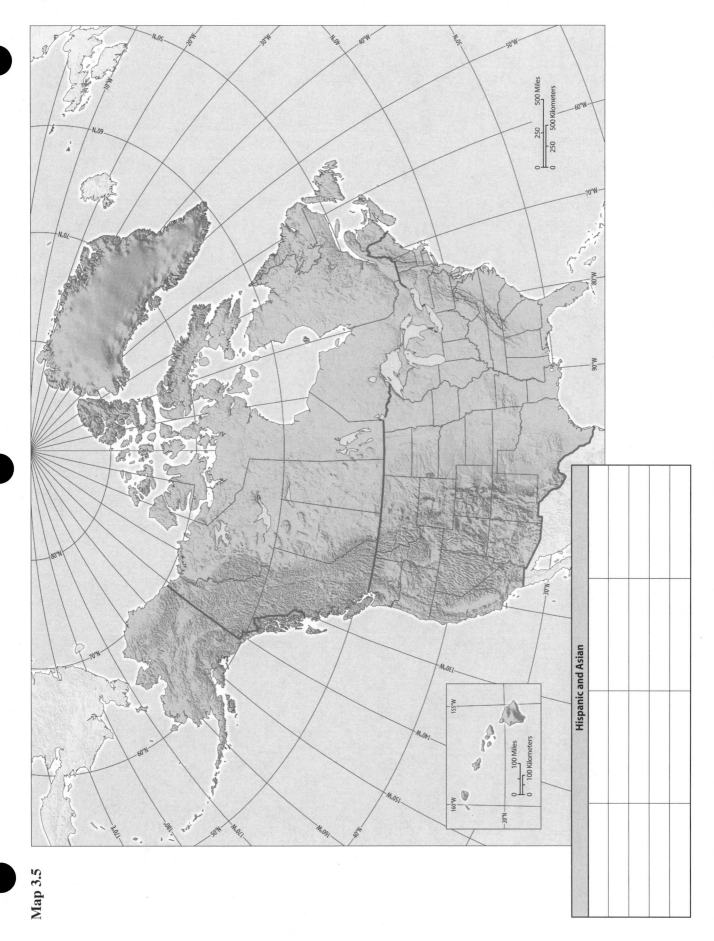

Hispanic and Asian

Map 3.5

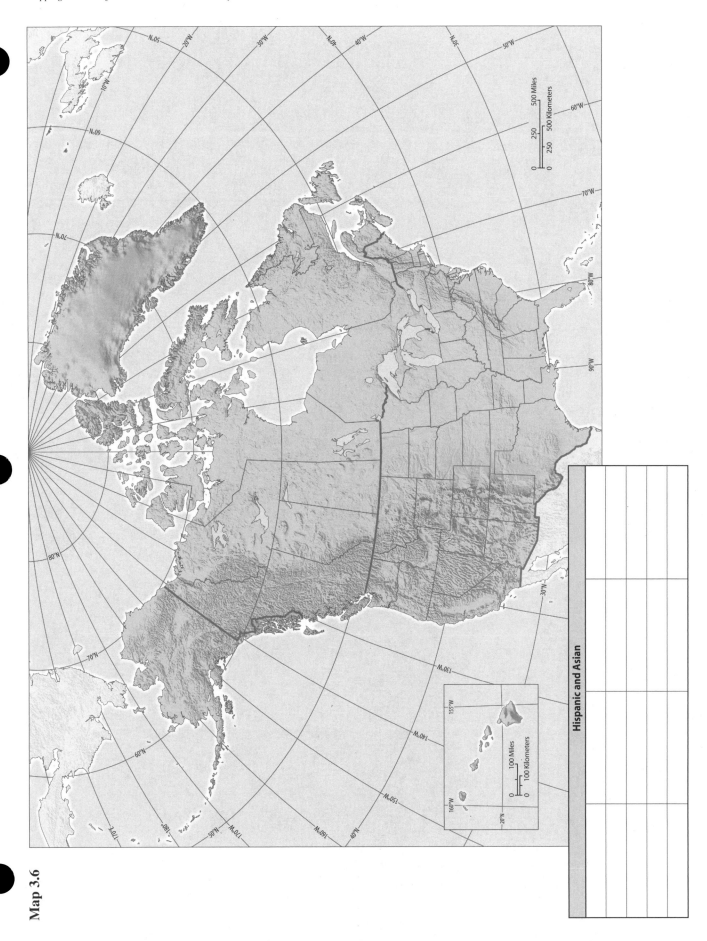

Map 3.6

Hispanic and Asian

Chapter Four: Latin America Mapping Workbook Exercises

Identify the following features on Mapping Workbook Maps 4.1, 4.2, and 4.3

Identify and label the following countries on Mapping Workbook Map 4.1

Argentina	Ecuador	Panama
Bolivia	El Salvador	Paraguay
Brazil	Guatemala	Peru
Chile	Honduras	Uruguay
Columbia	Mexico	Venezuela
Costa Rica	Nicaragua	

Identify and label the following cities on Mapping Workbook Map 4.2

Asunción	La Plata	Quito
Bogotá	Lima	Rio de Janeiro
Brasilia	Managua	San Jose
Buenos Aries	Manaus	San Salvador
Caracas	Medellin	Sal Paulo
Ciudad Juarez	Mexico City	Salvador
Cuzco	Monterrey	Santiago
Guatemala City	Montevideo	Sucre
Iquitos	Panama City	Tegucigalpa
La Paz	Punta Arenas	Tijuana

Identify and label the following physical features on Mapping Workbook Map 4.3

Amazon Basin	Guiana Highlands (Shield)	Rio de le Plata
Amazon River	Gulf of Mexico	San Francisco River
Andes Mountains	Llanos	Sierra Madre Occidental
Atlantic Ocean	Mato Grosso Plateau	Sierra Madre Oriental
Brazilian Highlands (Shield)	Negro River	Tierra del Fuego
Caribbean Sea	Orinoco River	Uruguay River
Cordillera Occidental	Pacific Ocean	Yucatan Peninsula
Cordillera Oriental	Parana River	
Falkland Islands	Patagonian Shield	
Galapagos Islands	Rio Bravo	

Map 4.1

Map 4.2

Map 4.3

Exercise One: Topography and Latin American Population Settlement

Using Figure 4.1 "Latin America" (physical geography) (p. 103), Figure 4.12 "Climate Map of Latin America" (p. 111), Figure 4.15 "Population Map of Latin America" (p. 114), and Mapping Workbook Map 4.4, complete the following exercise.

Using Figure 4.1 "Latin America" (physical geography) (p. 103) as a reference, use a colored pencil to shade in the low-elevation regions of Latin America (0–499 feet above sea level). Using a different colored pencil, shade in the higher-elevation regions of Latin America (500–1,999 feet above sea level). Using a different colored pencil, shade in the higher-elevation regions of Latin America (2,000–3,999 feet above sea level). Using a pencil that is a different color from the first two you used, shade in the highest elevation regions of Latin America (4,000+ feet above sea level). Mark the shading scheme in the map's legend.

Examine Figure 4.15 "Population Map of Latin America" (p. 114) and note the major settlement patterns. Using a red or black (or another color that will contrast with the shading scheme you have already created) colored pencil, shade in the Latin American regions possessing the highest population concentration. Mark the shading scheme in the map's legend. Once you have done this, answer the following questions.

1. Describe the general topographic pattern within Mexico and Central America.

_____

_____

_____

2. Describe the general topographic pattern within South America.

_____

_____

_____

3. Describe the general pattern of population concentration within Mexico and Central America.

_____

_____

_____

4. Describe the general pattern of population concentration within South America.

_____

_____

_____

5. Overall, how does the settlement and topography correlate within Mexico and Central America?

_____

_____

6. Overall, how does the settlement and topography correlate within South America?

_____

_____

_____

Examine the map you created and compare it with Figure 4.12 "Climate Map of Latin America" (p. 111).

7. In which climate(s) do we find the highest population density in Mexico and Central America?

_____

_____

8. Why do you think these regions are more densely inhabited?

_____

_____

_____

9. In which climate(s) do we find the lowest population density within Mexico and Central America?

_____

_____

10. Why do you think these regions are more sparsely inhabited?

_____

_____

_____

11. In which climate(s) do we find the highest population density in South America?

_____

_____

12. Why do you think these regions are more densely inhabited?

_____

_____

_____

13. In which climate(s) do we find the lowest population density within South America?

_____

_____

_____

14. Why do you think these regions are more sparsely inhabited?

_____

_____

Exercise Two: Latin American Economic Development

Using Table 4.2 "Development Indicators" (p. 128), Figure 4.32 "Global Linkages: Foreign Investment and Remittances" (p. 132), and Mapping Workbook Map 4.5, complete the following exercise.

Using the GNI per Capita, PPP 2010 column in Table 4.2 "Development Indicators" (p. 128) as a reference, use a colored pencil to shade in the Latin American countries possessing an annual per capita GNI of $2,790–$5,346. Using a different colored pencil, shade in the Latin American countries possessing an annual per capita GNI of $5,347–$7,902. Using a different colored pencil, shade in the Latin American countries possessing an annual per capita GNI of $7,903–$10,458. Using a different colored pencil, shade in the Latin American countries possessing an annual per capita GNI of $10,459–$13,014. Using a different colored pencil, shade in the Latin American countries possessing an annual per capita GNI of $13,015–$15,570. Mark your shading scheme in the map's legend.

Using Figure 4.32 "Global Linkages: Foreign Investment and Remittances" (p. 132) as a guide, take a ruler and draw proportional symbols representing the remittances per capita in U.S. dollars on their respective countries. Have a 1/2 inch (on one side) square represent greater than $300 remittances per capita, a 1/4 inch (on one side) square represent $201–$300 remittances per capita, a 1/8 inch (on one side) square represent $100–$200 remittances per capita, and a 1/16 inch (on one side) square represent less than $100 remittances per capita. Use a red or black (or another color that will contrast with the shading scheme you have created) colored pencil to shade in the proportional symbols on the map's legend.

Study the map you created and answer the following questions.

1. Which Latin American nations possess the highest annual per capita GNI?

_____

_____

2. Which Latin American nations possess the lowest annual per capita GNI?

_____

_____

3. Is there an overall regional pattern of annual per capita GNI for Latin America? If so, what is it?

_____

_____

Examine the remittances per capita proportional symbols you placed on your map.

4. Which nations receive the greatest remittances per capita?

_____

_____

5. Which nations receive the lowest remittances per capita?

_____

_____

6. Is there an overall regional pattern of remittances per capita for Latin America? If so, what is it (i.e., does the region display a sub-region that appears to receive a greater amount of remittances)?

_____

_____

_____

Compare and contrast the remittances per capita GNI data you mapped.

7. What is the correlation (positive or negative) between remittances and per capita GNI?

_____

_____

_____

8. If there was a correlation, what might explain it?

_____

_____

_____

9. If the distribution was more random, what might account for it?

_____

_____

_____

Table

GNI (US$)	Color
2,790–5,346	
5,347–7,902	
7,903–10,458	
10,459–13,014	
13,015–15,570	

Population	

Elevation	

Map 4.4

GNI	

Remittances Per Capita, 2009 (US $)	

Map 4.5

Chapter Five: The Caribbean Mapping Workbook Exercises

Identify the following features on Mapping Workbook Maps 5.1, 5.2, and 5.3

Identify and label the following countries on Mapping Workbook Map 5.1

Anguilla	Curacao	Martinique
Antigua and Barbuda	Dominica	Puerto Rico
Bahamas	Dominican Republic	St. Kitts and Nevis
Barbados	French Guiana	St. Lucia
Belize	Grenada	St. Vincent and the Grenadines
Bermuda	Guyana	Suriname
Bonaire	Haiti	Trinidad and Tobago
Cuba	Jamaica	

Identify and label the following cities on Mapping Workbook Map 5.2

Basse-Terre	Nassau
Belmopan	Paramaribo
Bridgetown	Port of Spain
Castries	Port-au-Prince
Cayenne	Roseau
Georgetown	San Juan
Havana	Santo Domingo
Kingston	St. George's
Kingstown	St. John's

Identify and label the following physical features on Mapping Workbook Map 5.3

Antillean Volcanic Arc
Atlantic Ocean
Caribbean Sea
Cordillera Central
Greater Antilles
Guiana Shield
Gulf of Mexico
Isla de Margarita
Leeward Islands
Lesser Antilles
Windward Islands
Windward Passage

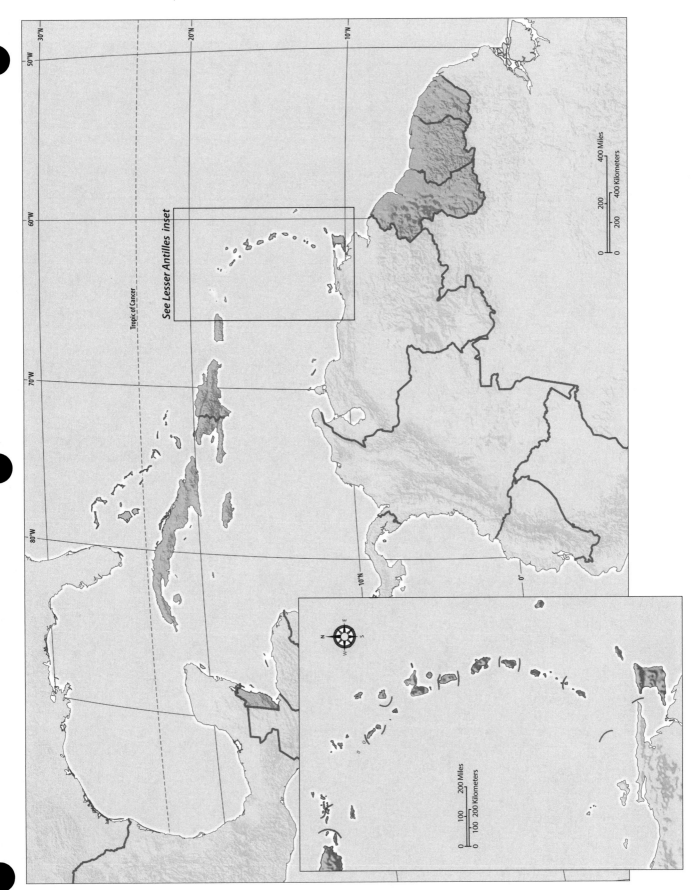

See Lesser Antilles inset

Map 5.1

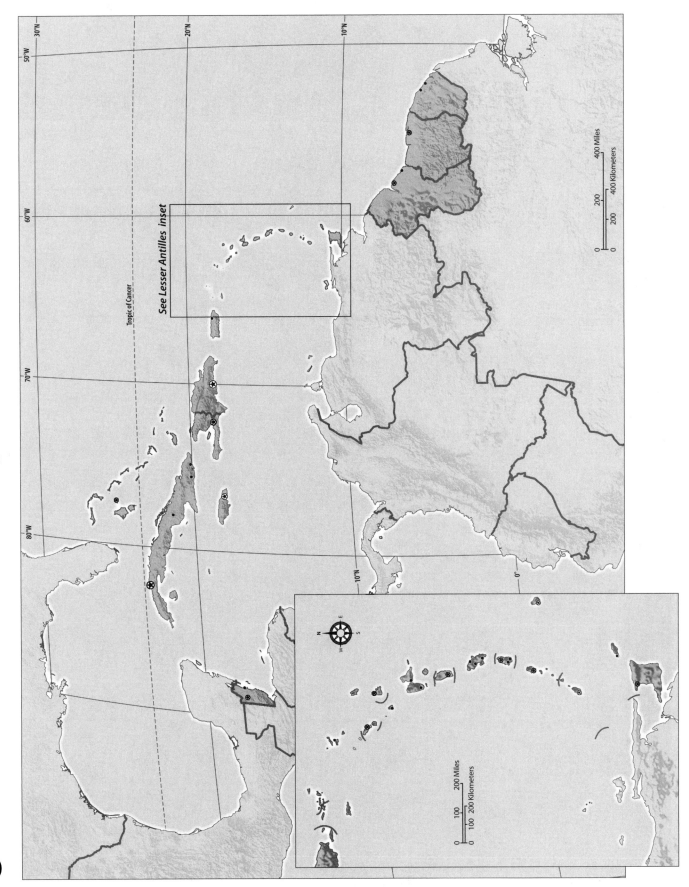

Map 5.2

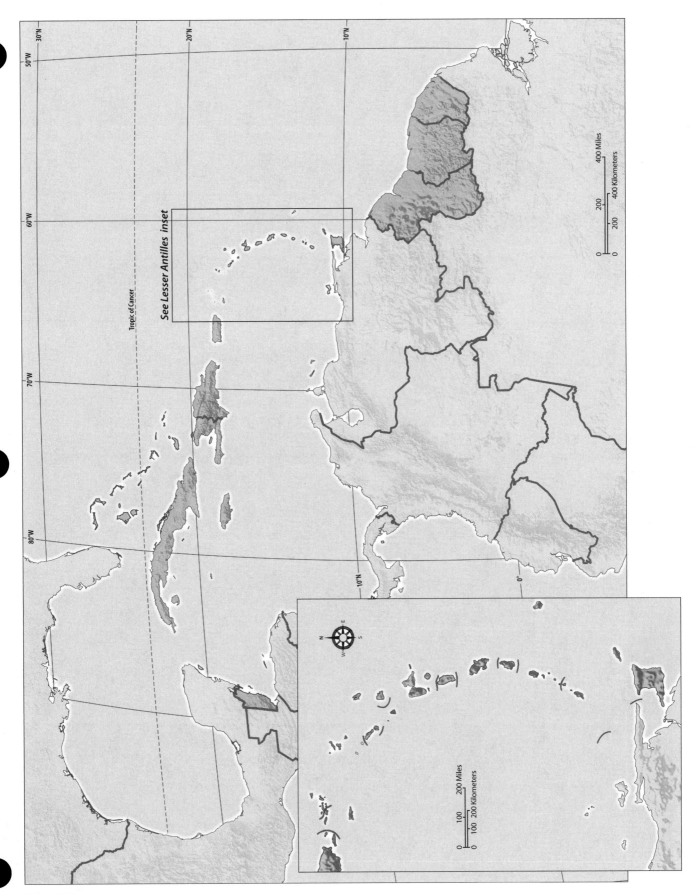

Map 5.3

Exercise One: Regional Patterns of the Caribbean Diaspora

Using Figure 5.14 "Caribbean Diaspora" (p. 151), Figure 5.21 "Caribbean Language Map" (p. 156), and Mapping Workbook Map 5.4, complete the following exercise.

Using Figure 5.21 "Caribbean Language Map" (p. 156) as a guide, use a colored pencil to shade in the Caribbean nations that possess Spanish as an official language. Using different colored pencils for each, shade in the nations possessing French, English, and Dutch as official languages. Mark your shading scheme in the map's legend. Once you do this, answer the following questions.

1. Where are the majority of the Spanish-speaking nations located within the region?

_____

_____

2. Where are the majority of the French-speaking nations located within the region?

_____

_____

3. Where are the majority of the English-speaking nations located within the region?

_____

_____

4. Where are the majority of the Dutch-speaking nations located within the region?

_____

_____

5. Based on your observations, can you identify a sub-regional pattern regarding these languages? If so, specify.

_____

_____

6. What are the factors associated with the dominance of certain European languages within the Caribbean?

_____

_____

7. Linguistically, how closely are these languages associated with their respective European counterparts? (Read the section on "Creolization and Caribbean Identity" on page 155 of your text.)

_____

_____

8. What other linguistic influences may have come into play in the evolution or stagnation of these languages? (Read the section on "Creolization and Caribbean Identity" on page 155 of your text.)

_____

_____

Examine Figure 5.14 "Caribbean Diaspora" (p. 151). Using a red or black (or another color that will contrast with the shading scheme you have created) colored pencil, draw flow arrows indicating the approximate origins and destinations of international and interregional migration within the Caribbean. Mark the arrow symbols in the map's legend. Once you have done this, do the following tasks and answer the questions that follow.

List the general destinations for migrants from the below Lesser Antilles, Greater Antilles, and Rimland nations (if it is outside the region depicted in Figure 5.14 "Caribbean Diaspora" [p. 151], just note the migration as "international").

9. Belize_____

10. French Guiana _____

11. Guyana _____

12. Suriname _____

13. Bahamas _____

14. Cuba _____

15. Dominican Republic _____

16. Haiti_____

17. Jamaica _____

18. Puerto Rico _____

19. Trinidad and Tobago _____

20. Smaller islands of the Lesser Antilles _____

21. Does there appear to be a geographic pattern associated with migration from the Rimland nations? If so, what is it?

_____

_____

22. What do you think may account for this pattern?

_____

_____

23. Does there appear to be a geographic pattern associated with migration from the Greater Antilles nations? If so, what is it?

_____

_____

24. What do you think may account for this pattern?

_____

_____

25. Does there appear to be a geographic pattern associated with migration from the Lesser Antilles nations? If so, what is it?

_____

_____

26. What do you think may account for this pattern?

_____

_____

Examine the two mapped variables (language and migration) together.

27. Does there appear to be a correlation between migration origins, destinations, and language? If so, describe it?

_____

_____

28. If there was a correlation, how do you explain it?

_____

_____

Exercise Two: Caribbean Economic Development

Using Table 5.2 "Development Indicators" (p. 162), Figure 5.31 "Development Issues: Human Development and Remittances in the Caribbean" (p. 167), and Mapping Workbook Map 5.5, complete the following exercise.

Using the GNI per Capita, PPP 2010, column in Table 5.2 "Development Indicators" (p. 162) as a reference, use a colored pencil to shade in the Caribbean countries possessing an annual per capita GNI of $1,180–$5,904. Using a different colored pencil, shade in the Caribbean countries possessing an annual per capita GNI of $5,905–$10,628. Using a different colored pencil, shade in the Caribbean countries possessing an annual per capita GNI of $10,629–$15,352. Using a different colored pencil, shade in the Caribbean countries possessing an annual per capita GNI of $15,353–$20,076. Using a different colored pencil, shade in the Caribbean countries possessing an annual per capita GNI of $20,077 and above. Mark your shading scheme in the map's legend. Once you have done this, answer the following questions.

1. Which Caribbean and Rimland nations possess the highest annual per capita GNI?

_____

_____

2. Which Caribbean and Rimland nations possess the lowest annual per capita GNI?

_____

_____

3. Is there an overall regional pattern of annual per capita GNI for the Caribbean and the Rimland? If so, what is it (i.e., does the region display a more and a less impoverished sub-region)?

_____

_____

_____

4. Which sub-region appears to be the wealthiest, if any: the Greater Antilles, the Lesser Antilles, or the Rimland?

_____

_____

5. Can you explain the patterns of economic development, or lack thereof, in the Caribbean?

_____

_____

_____

Using Figure 5.31 "Development Issues: Human Development and Remittances in the Caribbean" (p. 167), as a guide, take a ruler and draw respective proportional symbols representing annual remittances per capita in U.S. dollars on their respective countries. Have a 1/2 inch (on one side) square represent $700 and above per capita per year remittances, a 1/4 inch (on one side) square represent $500 to $699 per capita per year remittances, a 1/8 inch (on one side) square represent $300 to $499 per capita per year remittances, and a 1/16 inch (on one side) square represent less than $499 per capita per year remittances. Use a red or black colored pencil (or another color that will contrast with the shading scheme you have created) to shade in the proportional symbols below.

Examine the tourist dollar receipts proportional symbols you placed on your map.

6. Which nations receive the greatest tourist dollar receipts?

_____

_____

7. Which nations receive the lowest tourist dollar receipts?

_____

_____

8. Is there an overall regional pattern of tourist dollar receipts for the Caribbean? If so, what is it (e.g., does the region display a sub-region that appears to receive a greater amount of tourist dollars)?

_____

_____

_____

Compare and contrast the tourist dollars with the annual per capita GNI data you mapped.

9. Is there a correlation (positive or negative) between tourist dollars and per capita GNI? If so, what is it? If so, what might explain it?

_____

_____

_____

_____

_____

_____

GNI (US$)	Color
1,180–5,904	
5,905–10,628	
10,629–15,352	
15,353–20,076	
20,077 and above	

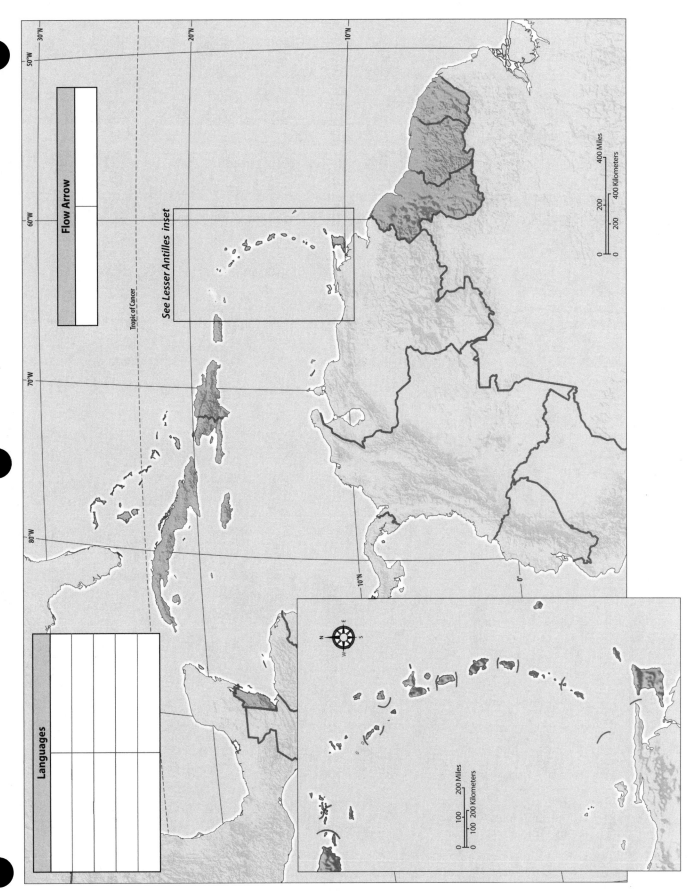

Flow Arrow

Tropic of Cancer

See Lesser Antilles inset

Languages

50°W 60°W 70°W 80°W

30°N 20°N 10°N 0°

10°N

400 Miles
400 Kilometers
200
200
0
0

200 Miles
100 200 Kilometers
100
0
0

N
E
W
S

Map 5.4

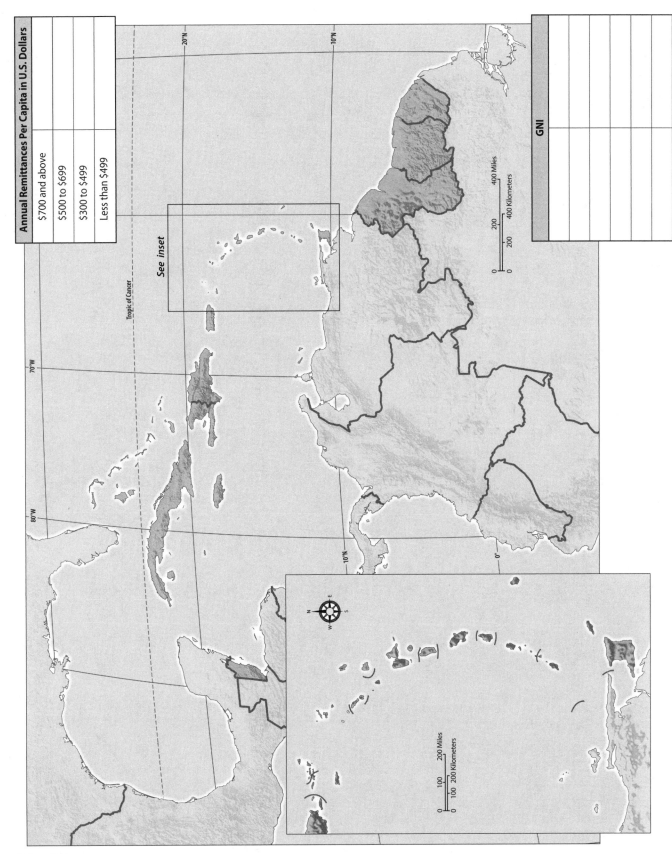

Annual Remittances Per Capita in U.S. Dollars		
$700 and above		
$500 to $699		
$300 to $499		
Less than $499		

See inset

Tropic of Cancer

20°N

10°N

70°W

80°W

10°N

0°

GNI					

0   200   400 Miles

0   200   400 Kilometers

0   100   200 Miles

0   100   200 Kilometers

Map 5.5

Chapter Six: Sub-Saharan Africa Mapping Workbook Exercises

Identify the following features on Mapping Workbook Maps 6.1, 6.2, and 6.3

Identify and label the following countries on Mapping Workbook Map 6.1

Angola	Gambia	Rwanda
Benin	Ghana	Sao Tome and Principe
Botswana	Guinea	Senegal
Burkina Faso	Guinea-Bissau	Seychelles
Burundi	Ivory Coast	Sierra Leone
Cabinda (Angola)	Kenya	Somalia
Cameroon	Lesotho	South Africa
Cape Verde	Liberia	South Sudan
Central African Republic	Madagascar	Sudan
Chad	Malawi	Swaziland
Comoros Islands	Mali	Tanzania
Democratic Republic of the Congo	Mauritania	Togo
Djibouti	Mauritius	Uganda
Equatorial Guinea	Mozambique	Western Sahara
Eritrea	Namibia	Zambia
Ethiopia	Niger	Zimbabwe
Gabon	Nigeria	
	Republic of the Congo	

Identify and label the following cities on Mapping Workbook Map 6.2

Abuja	Harare	N'Djamena
Accra	Johannesburg	Nairobi
Addis Ababa	Kampala	Niamey
Antananarivo	Khartoum	Nouakchott
Asmara	Kigali	Ouagadougou
Bamako	Kinshasa	Port Louis
Bangui	Lagos	Porto-Novo
Banjul	Libreville	Praia
Bissau	Lilongwe	Pretoria
Brazzaville	Lome	Sao Tome
Bujumbura	Luanda	Tombocutou
Cape Town	Lusaka	Victoria
Conakry	Malabo	Windhoek
Dakar	Maputo	Yamoussoukro
Dar es Salaam	Maseru	Yaounde
Djibouti	Mogadishu	
Freetown	Monrovia	
Gaborone	Moroni	

Identify and label the following physical features on Mapping Workbook Map 6.3

Adamawa Highlands	Horn of Africa	Namib Desert
Atlantic Ocean	Indian Ocean	Niger River
Benue River	Kalahari Desert	Nile River
Blue Nile River	Kasai River	Orange River
Cape of Good Hope	Lake Chad	Red Sea
Caprivi Strip	Lake Nyasa	Reunion
Congo Basin	Lake Tana	Sahara Desert
Congo River	Lake Tanganyika	Sahel
Ethiopian Highlands	Lake Turkana	Senegal River
Gambia River	Lake Victoria	Ubangi River
Great Escarpment	Limpopo River	White Nile River
Great Rift Valley	Mozambique Channel	Zambezi River
Gulf of Aden	Mt. Kenya	
Gulf of Guinea	Mt. Kilimanjaro	

Map 6.1

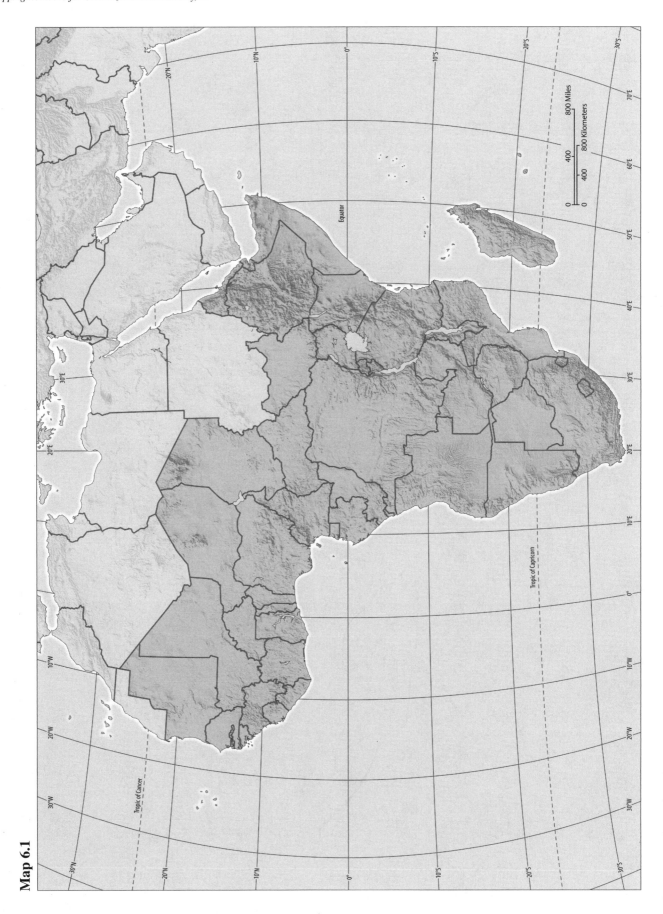

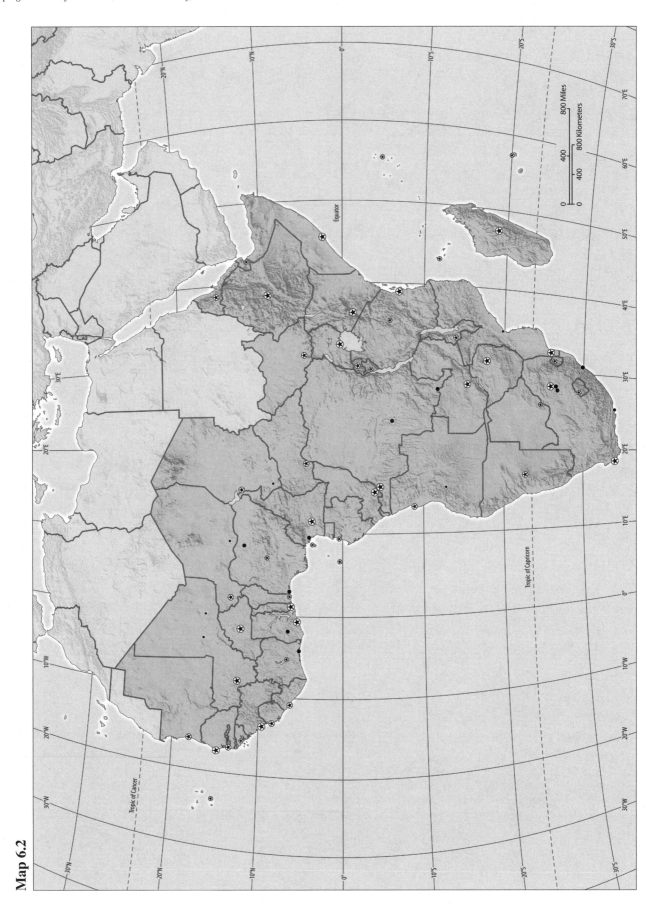

Map 6.2

Map 6.3

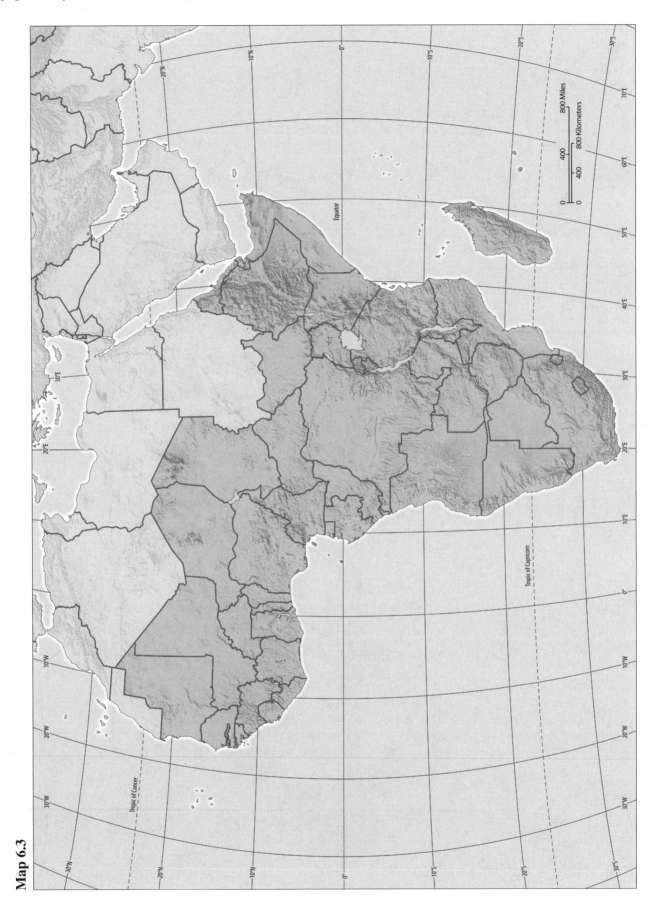

Exercise One: Regional Patterns of the Language and Colonization

Using Figure 6.24 Language Map of Sub-Saharan Africa: Language Groups and Official Languages (p. 191), Figure 6.33 European Colonization in 1913 (p. 199), and Mapping Workbook Map 6.4, complete the following exercise.

Using Figure 6.24 "Language Map of Sub-Saharan Africa: Language Groups and Official Languages" (p. 191) as a reference, use a colored pencil to shade in the African nations that have Amharic as their official language. Using a different colored pencil to represent each language, shade in the official languages of each of the African nations. Mark your shading scheme in the map's legend. Once you have done this, answer the following questions.

1. In which Sub-Saharan African nation(s) do you find Amharic as the official language?

_____

_____

2. In which Sub-Saharan African nation(s) do you find Arabic as the official language?

_____

_____

3. In which Sub-Saharan African nation(s) do you find English as the official language?

_____

_____

4. In which Sub-Saharan African nation(s) do you find French as the official language?

_____

_____

5. In which Sub-Saharan African nation(s) do you find Portuguese as the official language?

_____

_____

6. In which Sub-Saharan African nation(s) do you find Somali as the official language?

_____

_____

7. In which Sub-Saharan African nation(s) do you find Spanish as the official language?

_____

_____

8. In which Sub-Saharan African nation(s) do you find Swahili as the official language?

_____

_____

9. In which Sub-Saharan African nation(s) do you find no official language or a dominance of multiple languages?

_____

_____

10. Does there appear to be a regional pattern associated with the languages within Sub-Saharan Africa? If so, what is it?

_____

_____

_____

11. If there is a regional pattern, what do you think the cause of it is?

_____

_____

_____

Examine Figure 6.33 "European Colonization in 1913" (p. 199). Draw a 1/4 inch square over each Sub-Saharan African nation. Use a colored pencil to pattern each square to represent each of the major European colonizing nations that are listed in the legend of Figure 6.33 "European Colonization in 1913" (p. 199) (try to select a color that will contrast with the colors you have used to depict official languages). Mark your patterning scheme in the map's legend. Once you have done this, answer the questions below.

Colonizing nation	Symbol	Colonizing nation	Symbol
Belgium		Italy	
Britain		Portugal	
France		Spain	
Germany		Independent	

12. Which Sub-Saharan African nation(s) today have a Belgian colonial legacy?

_____

_____

13. Which Sub-Saharan African nation(s) today have a British colonial legacy?

_____

_____

14. Which Sub-Saharan African nation(s) today have a French colonial legacy?

_____

_____

15. Which Sub-Saharan African nation(s) today have a German colonial legacy?

_____

_____

16. Which Sub-Saharan African nation(s) today have an Italian colonial legacy?

_____

_____

17. Which Sub-Saharan African nation(s) today have a Portuguese colonial legacy?

_____

_____

18. Which Sub-Saharan African nation(s) today have a Spanish colonial legacy?

_____

_____

19. Which Sub-Saharan African nation(s) have never been colonized?

_____

_____

20. Is there a regional pattern associated with European colonization? If so, what is it?

_____

_____

_____

_____

21. How would you explain the pattern of colonization?

_____

_____

_____

_____

_____

Compare and contrast the official languages within the various Sub-Saharan African nations with the colonial legacy symbols you drew.

22. Is there a correlation between the European colonizers and official languages within these nations? Please provide specific examples.

_____

_____

_____

_____

_____

23. What do you believe to be the cause of this correlation?

24. Are there any outliers to this situation? For example, are there Sub-Saharan African nations that possess a completely different official language from that of their colonizers? Please provide specific examples.

25. What do you think are the causes of these outliers?

Exercise Two: Patterns of Change Under Age 5 Mortality in Sub-Saharan Africa

Using Table 6.2 Development Indicators (p. 204) and Mapping Workbook Map 6.5, complete the following exercise.

Examine Table 6.2 "Development Indicators" (p. 204) and in the table below record the under age 5 mortality rates for 1990 and 2010 in their respective columns. Once you do this, determine the difference between the 1990 and 2010 rates by subtracting the 2010 under age 5 mortality rates from the 1990 under age 5 mortality rates. Record the difference in the "Rate Difference" column in the table. Be certain to indicate whether there was a decrease (use a minus "–" symbol to indicate a decrease) or an increase (use a "+" symbol to indicate an increase) in the under age 5 mortality rate.

Once you have done this, use a colored pencil to shade in those countries that have experienced a *decrease* in their under age 5 mortality rates. Specifically, use one color to indicate countries that have experienced a net decrease of 168–134 in their under age 5 mortality rate, a different colored pencil to indicate countries that have experienced a net decrease of 133–101 in their under age 5 mortality rate, a different colored pencil to indicate countries that have experienced a net decrease of 100–67 in their under age 5 mortality rate, a different colored pencil to indicate countries that have experienced a net decrease of 66–34 in their under age 5 mortality rate, and a different colored pencil to indicate countries that have experienced a net decrease of 33–1 in their under age 5 mortality rate. Also, shade in the respective colors in the legend below.

Rate of decrease under age 5 mortality	Color
–168–134	
–133–101	
–100–67	
–66–34	
–33–1	

Once you have done this, answer the following questions.

1. Which Sub-Saharan African nations have experienced the largest rates of decrease in their under age 5 mortality between 1990 and 2010?

_____

_____

_____

2. Does there appear to be a geographic pattern associated with those Sub-Saharan African nations that have experienced such a decrease?

_____

_____

_____

3. If there is a geographic pattern, what could possibly explain it?

_____

_____

_____

_____

_____

_____

4. Closely examine the data for the nations that have experienced the lowest rates of change in the decrease of their under age 5 mortality (not those that have experienced a gain, but rather those that have experienced rates of only –33–1) between 1990 and 2010. In regard to these nations, what can be stated about their lower rates of decrease?

_____

_____

_____

_____

_____

_____

5. Now closely examine the data of those nations that have experienced the highest rate of decrease (–168–134) in their under age 5 mortality between 1990 and 2010. In regard to these nations, what can be stated about their higher rates of decrease?

_____

_____

_____

_____

_____

_____

6. Now contrast the data of the nations that experienced the lowest rate of decrease (–33–1) with the data of those nations that experienced the greatest rate of decrease (–168–134) in under age 5 mortality between 1990 and 2010. Is a small decrease in the under age 5 mortality in a given nation as compared to a large decrease in the under age 5 mortality rate in a given nation a comparatively better situation? Explain.

_____

_____

_____

_____

_____

7. Which Sub-Saharan African nation(s) experienced an overall increase in their rates of under age 5 mortality between 1990 and 2010?

_____

_____

_____

8. Is there a geographic pattern associated with those Sub-Saharan African nations that have experienced such an increase, and if so, what could explain this pattern?

_____

_____

_____

9. What could be the possible causes of increased rates of under age 5 mortality within the nation(s) that experienced an increase in its rate of under age 5 mortality during this period?

_____

_____

_____

_____

_____

_____

_____

_____

_____

Country	Under age 5 mortality rate, 1990	Under age 5 mortality rate, 2010	Rate difference (= 1990 – 2010 data)
Angola			
Benin			
Botswana			
Burkina Faso			
Burundi			
Cameroon			
Cape Verde			
Central African Republic			
Chad			
Comoros			
Congo			
Democratic Republic of the Congo			
Djibouti			
Equatorial Guinea			
Eritrea			
Ethiopia			
Gabon			
Gambia			
Ghana			
Guinea			

Country	Under age 5 mortality rate, 1990	Under age 5 mortality rate, 2010	Rate difference (= 1990 – 2010 data)
Guinea-Bissau			
Ivory Coast			
Kenya			
Lesotho			
Liberia			
Madagascar			
Malawi			
Mali			
Mauritania			
Mauritius			
Mozambique			
Namibia			
Niger			
Nigeria			
Rwanda			
Sao Tome and Principe			
Senegal			
Seychelles			
Sierra Leone			
Somalia			
South Africa			
South Sudan			
Swaziland			
Tanzania			
Togo			
Uganda			
Zambia			
Zimbabwe			

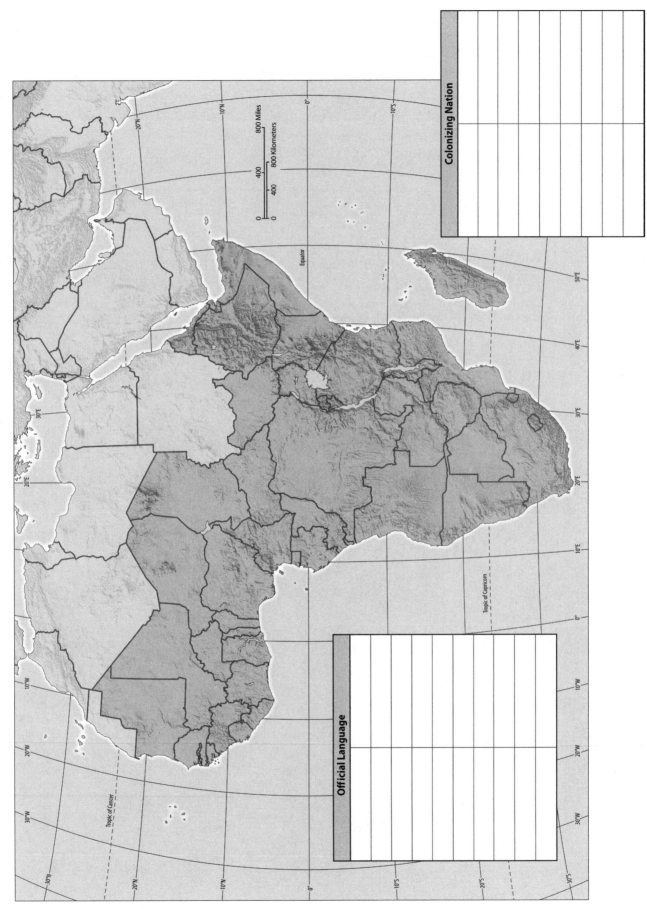

Colonizing Nation

Official Language

Map 6.4

Map 6.5

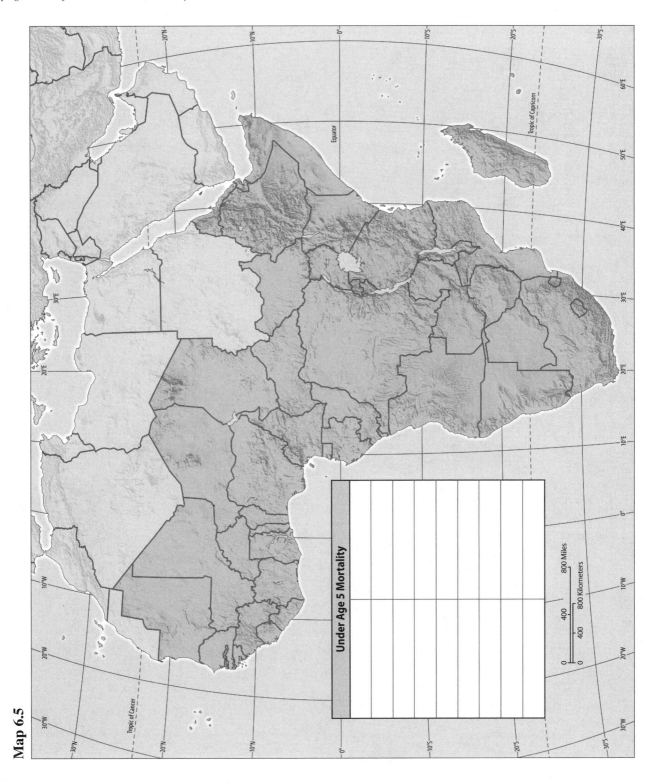

Under Age 5 Mortality

800 Miles

800 Kilometers

400

400

0

0

Tropic of Capricorn

Equator

Tropic of Cancer

Chapter Seven: Southwest Asia and North Africa Mapping Workbook Exercises

Identify the following features on Mapping Workbook Maps 7.1, 7.2, and 7.3

Identify and label the following countries on Mapping Workbook Map 7.1

Algeria	Kuwait	Sudan
Bahrain	Lebanon	Syria
Egypt	Libya	Tunisia
Iran	Morocco	Turkey
Iraq	Oman	United Arab Emirates
Israel	Qatar	Western Sahara
Jordan	Saudi Arabia	Yemen

Identify and label the following cities on Mapping Workbook Map 7.2

Abu Dhabi	Doha	Muscat
Algiers	Dubai	Rabat
Amman	El Aaiun	Riyadh
Ankara	Istanbul	San'a
Baghdad	Jerusalem	Tehran
Basra	Kuwait City	Tel Aviv
Beirut	Madinah	Tripoli
Bursa	Manama	Tunis
Cairo	Makkah	
Damascus	Mosul	

Identify and label the following physical features on Mapping Workbook Map 7.3

Ahaggar Highlands	Indian Ocean	Rub-Al-Khali
Anatolian Plateau	Iranian Plateau	Sahara Desert
Arabian Peninsula	Jordan River	Sinai
Atlantic Ocean	Lake Nasser	Socotra
Atlas Mountains	Libyan Desert	Strait of Gibraltar
Black Sea	Maghreb	Straits of Hormuz
Blue Nile River	Mediterranean Sea	Suez Canal
Caspian Sea	Nile Delta	Tigris River
Dead Sea	Nile River	White Nile River
Elburz Mountains	Nubian Desert	Yemen Highlands
Euphrates River	Persian Gulf	Zagros Mountains
Gulf of Aden	Red Sea	

Map 7.1

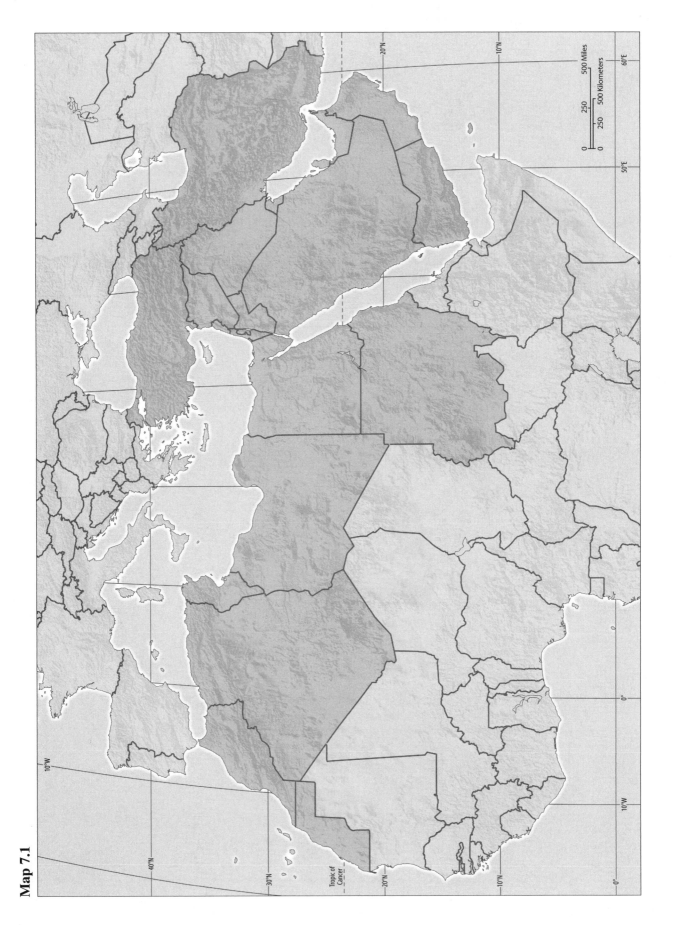

Map 7.2

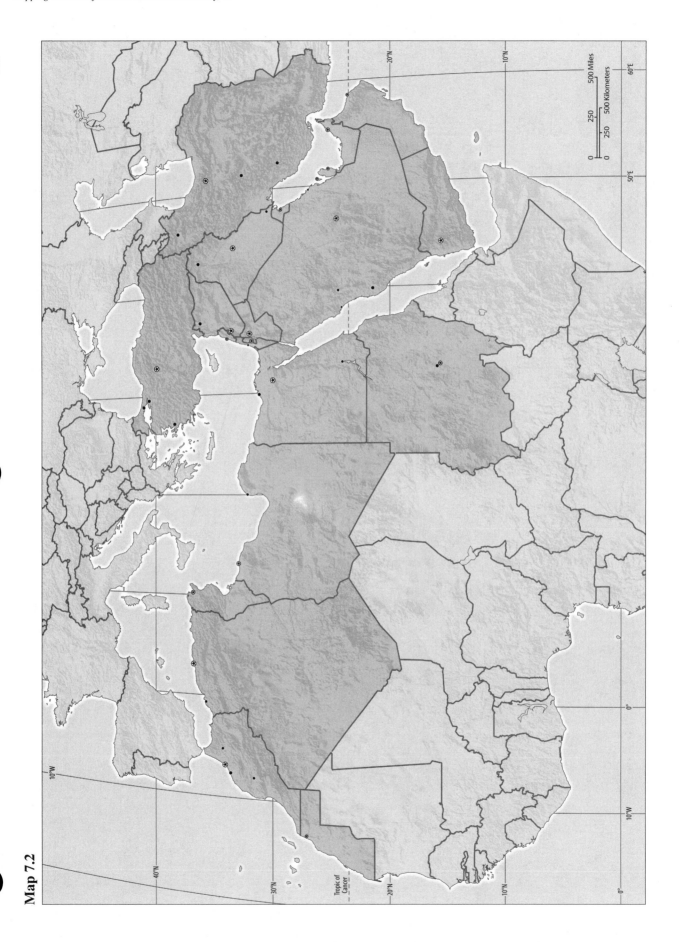

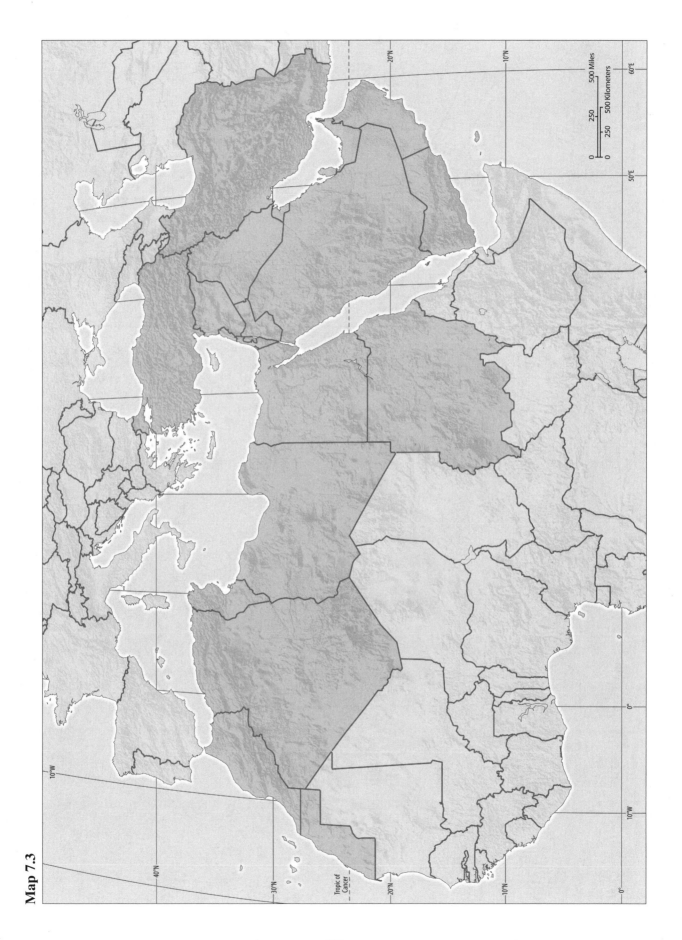

Map 7.3

Exercise One: Climate and Southwest Asian and North African Agriculture

Using Figure 7.1 "Southwest Asia and North Africa" (physical features) (pp. 216–217), Figure 7.11 "Climate Map of Southwest Asia and North Africa" (p. 223), Figure 7.17 "Agricultural Regions of Southwest Asia and North Africa" (p. 227), and Mapping Workbook Map 7.4, complete the following exercise.

Using Figure 7.11 "Climate Map of Southwest Asia and North Africa" (p. 223) as a reference, use a colored pencil to shade in the regions that possess a BWh climate. Using a different color to represent each climate type, shade in the other major climates within the region. Mark your shading scheme in the map's legend. Once you have done this, answer the following questions.

1. What are the major climates of Southwest Asia and North Africa?

_____

_____

2. What are the general locations of these climates?

_____

_____

Using Figure 7.1 "Southwest Asia and North Africa" (physical features) (pp. 216–217) as a reference, use a colored pencil to draw symbols representing mountains, deserts, rivers, and other physical features. Use different symbols for each category of feature. Mark the symbols in the map's legend. Once you have done this, answer the following questions.

3. What are the major physical features of Southwest Asia and North Africa?

_____

_____

4. What are the general locations of these physical features?

_____

_____

Examine the physical features you drew and compare them with the major climates that you shaded in on your map.

5. Do you see a pattern between the location of some of these physical features and the climate types within Southwest Asia and North Africa? If so, what are they? Please provide specific examples.

_____

_____

_____

_____

6. How is the climate affected by some of these physical features? More specifically, what are some of the controls that the physical features within this region have on climate? Explain.

_____

_____

_____

_____

Using Figure 7.17 "Agricultural Regions of Southwest Asia and North Africa" (p. 227) as a reference, use a red or black (or another color that contrasts with the shading scheme you created for climate) colored pencil to shade in the major agricultural types within the region. Use different patterns for pastoral nomadism, oasis and irrigated agriculture, and dry farming (with some irrigation). Mark this shading/patterning scheme in the map's legend. Once you have done this, answer the following questions.

7. What is (are) the location(s) of pastoral nomadism within Southwest Asia and North Africa?

_____

_____

8. What is (are) the location(s) of oasis and irrigated agriculture within Southwest Asia and North Africa?

_____

_____

9. What is (are) the location(s) of dry farming (with some irrigation) within Southwest Asia and North Africa?

_____

_____

Examine the agricultural regions and their location relative to the climate types and physical features you shaded in on the map.

10. What physical features and climate correlate with pastoral nomadism, if any?

_____

_____

11. How would you explain the pattern between this agricultural activity and the physical features and climate within the region where it prevails? If there was no correlation, how would you explain the lack of one?

_____

_____

12. What physical features and climate correlate with oasis and irrigated agriculture, if any?

_____

_____

13. How would you explain the pattern between this agricultural activity and the physical features and climate within the region where it prevails? If there was no correlation, how would you explain the lack of one?

_____

_____

14. What physical features and climate correlate with dry farming (with some irrigation), if any?

_____

_____

15. How would you explain the pattern between this agricultural activity and the physical features and climate within the region where it prevails? If there was no correlation, how would you explain the lack of one?

_____

_____

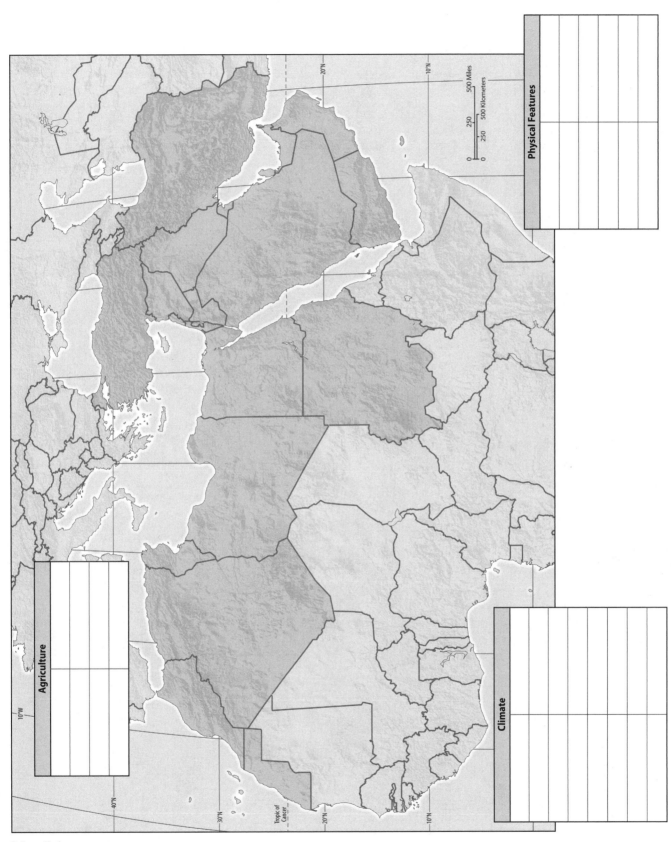

Map 7.4

Exercise Two: Changing Political Borders Within Israel

Using Figure 7.32 "Evolution of Israel" (p. 240), Figure 7.33 "West Bank" (p. 241), Figure 7.35 "Israeli Security Barrier" (p. 242), and Mapping Workbook Map 7.5, complete the following exercise.

Using Figure 7.32 "Evolution of Israel" (p. 240) as a reference, use a colored pencil to draw a line around the region that was established by the British between 1922 and 1948. Label this line "1922–1948." Using different colored pencils to represent each isochron (line of equal time), draw lines depicting the evolution of Israel under the UN partitioning plan of 1947, Israel after 1948–1949, and the region since the 1967 war. Label each line with its respective date. Mark these lines in the map's legend. Once you have done this, answer the following questions.

1. Describe the relative areas and comparative geographic locations of the Arab Muslim and Jewish states that were created under the UN partitioning plan.

_____

_____

_____

_____

2. Generally describe the circumstances that brought about these territorial changes.

_____

_____

_____

_____

3. How might have the size and the locations of these states had an effect on the access to resources and on the political relations between the citizens of these two states?

_____

_____

_____

4. Describe the relative areas and comparative geographic locations of Israel and Jordan after 1948–1949.

_____

_____

_____

_____

5. Generally describe the circumstances that brought about these territorial changes.

_____

_____

_____

_____

6. How might have the size and the locations of these states had an effect on the access to resources and on the political relations between the citizens of these two states, and with other inhabitants within the bordering nations?

_____

_____

_____

7. Describe the relative areas and comparative geographic locations of Israel, Jordan, Egypt, and Lebanon after the 1967 war.

_____

_____

_____

_____

8. Generally describe the circumstances that brought about these territorial changes.

_____

_____

_____

_____

9. How might have the size and the locations of these states had an effect on the access to resources and on the political relations between the citizens of these two states, and with other inhabitants within the bordering nations?

_____

_____

_____

Using Figure 7.33 "West Bank" (p. 241) as a reference, use a red or black colored pencil to draw in the general location of West Bank Palestinian settlements that are under full or partial Palestinian control. Mark this symbol in the map's legend. Once you have done this, answer the following questions.

10. Describe the general location of the Palestinian settlements within the West Bank. Are they clustered, dispersed, or in another pattern?

_____

_____

_____

11. How do you think the locations of these settlements have affected the political success and unity of the Palestinians within the West Bank? Explain.

_____

_____

_____

_____

Using Figure 7.35 "Israeli Security Barrier" (p. 242) as a reference, use a colored pencil (one that will contrast with the other colors you have already used) to draw a line indicating the location of the security barrier and the general location of the West Bank Israeli settlements. Mark this symbol in the map's legend. Once you have done this, answer the following questions.

12. Describe the general location of the Israeli settlements within the West Bank. Are they clustered, dispersed, or in another pattern?

_____

_____

_____

_____

13. How do you think the locations of these settlements have affected the political success and unity of the Israelis within the West Bank? Explain.

_____

_____

_____

_____

14. Compare the location of the Palestinian West Bank settlements and the location of the Israeli West Bank settlements. Describe the relative locations of these settlements.

_____

_____

_____

_____

_____

_____

15. How might their locations affect the political success and unity of the Palestinians and Israelis within the West Bank?

_____

_____

_____

16. Examine the security barrier line you drew on your map. How well does the line correspond to the locations of the Palestinian and Israeli settlements within the West Bank? Specifically, describe its relative location to these settlements.

_____

_____

_____

_____

_____

_____

17. Considering such issues as the movement of people and goods into and out of, as well as within, the West Bank, how might this security barrier affect these settlements and their inhabitants within the region?

_____

_____

_____

_____

18. How might the continued construction of the settlement barrier affect the political situation within the region?

_____

_____

_____

_____

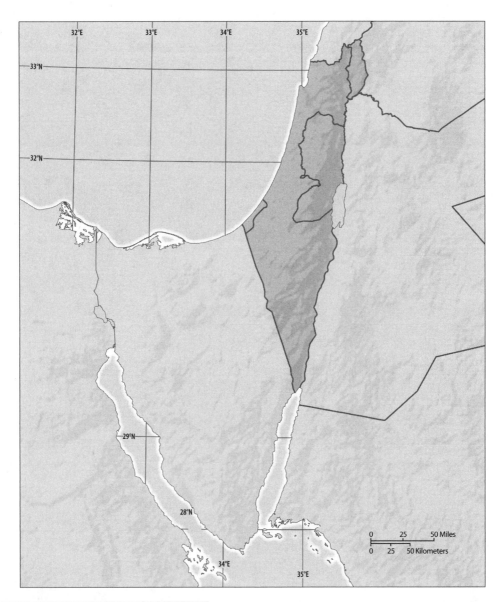

Isochron	

Palestinian Settlement and Security Barrier	

Map 7.5

Chapter Eight: Europe Mapping Workbook Exercises

Identify the following features on Mapping Workbook Maps 8.1, 8.2, and 8.3

Identify and label the following countries on Mapping Workbook Map 8.1

Albania	Greece	Netherlands
Andorra	Hungary	Norway
Austria	Iceland	Poland
Belgium	Ireland	Portugal
Bosnia and Herzegovina	Italy	Romania
Bulgaria	Kosovo	San Marino
Croatia	Latvia	Serbia
Cyprus	Liechtenstein	Slovakia
Czech Republic	Lithuania	Slovenia
Denmark	Luxembourg	Spain
Estonia	Macedonia	Sweden
Finland	Malta	Switzerland
France	Monaco	United Kingdom
Germany	Montenegro	Vatican City

Identify and label the following cities on Mapping Workbook Map 8.2

Amsterdam	Lisbon	Rome
Athens	Ljubljana	Sarajevo
Belfast	London	Skopje
Belgrade	Luxembourg	Sofia
Berlin	Madrid	Stockholm
Bern	Nicosia	Tallinn
Bratislava	Oslo	Tirana
Brussels	Palermo	Valletta
Bucharest	Paris	Vienna
Budapest	Podgorica	Vilnius
Copenhagen	Prague	Warsaw
Dublin	Pristina	Zagreb
Glasgow	Reykjavik	
Helsinki	Riga	

Identify and label the following physical features on Mapping Workbook Map 8.3

Adriatic Sea	Ebro River	Pyrenees
Aegean Sea	Elbe River	Rhine River
Alps	English Channel	Rhone River
Appenines Mountains	Faroe Islands	Sardinia
Atlantic Ocean	Garonne River	Seine River
Balearic Islands	Loire River	Shetland Islands
Baltic Sea	Mediterranean Sea	Sicily
Bay of Biscay	North Sea	Strait of Gibraltar
Carpathians	North European Lowland	Tagus River
Corsica	Norwegian Sea	Thames River
Crete	Oder River	Vistula River
Danube River	Orkney Islands	
Dinaric Alps	Po River	

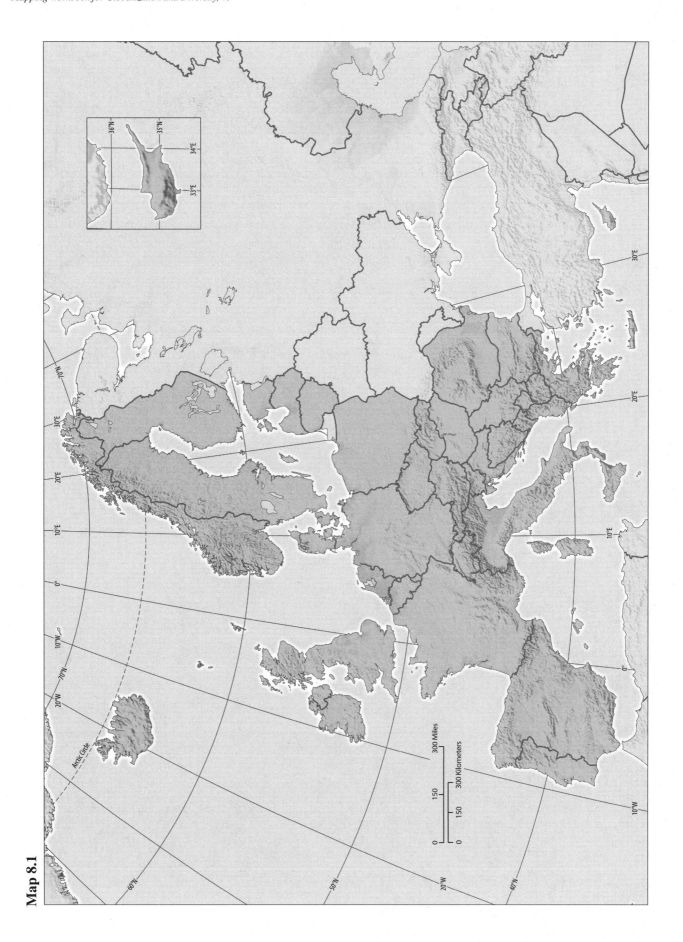

Map 8.1

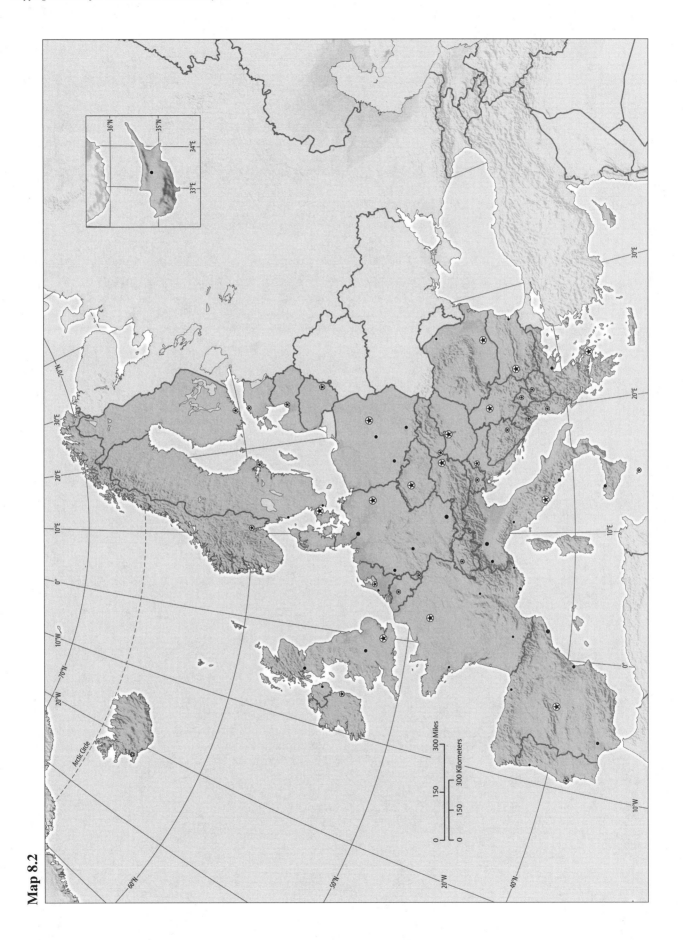

Map 8.2

Map 8.3

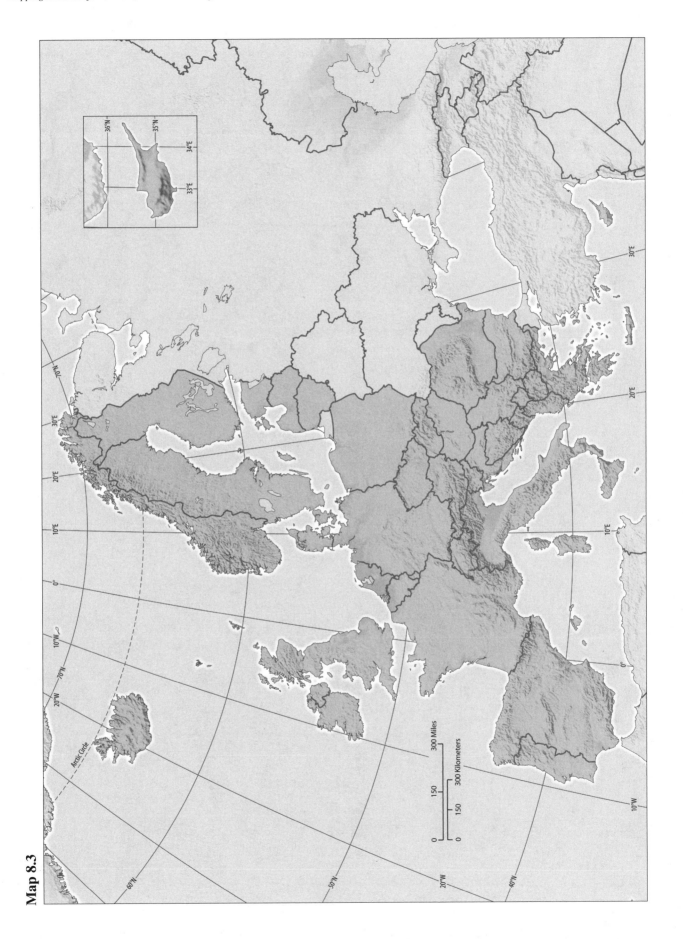

Exercise One: Regional Patterns of Pollution and Industrialization

Using Figure 8.8 "Environmental Issues in Europe" (p. 259), Figure 8.31 "Industrial Regions of Europe" (p. 282), and Mapping Workbook Map 8.4, complete the following exercise.

Using Figure 8.8 "Environmental Issues in Europe" (p. 259) as a reference, use a colored pencil to shade in the European regions that have been affected by acid rain. Mark this shading or scheme on the map's legend. Using a different colored pencil, shade in the major European rivers that have been polluted. Mark this symbol on the map's legend. Once you have done this, answer the following questions.

1. Define acid rain.

_____

_____

2. What are the major causes of acid rain?

_____

_____

3. Describe the extent of acid rain damage in Europe. Specifically, which European countries have been greatly affected by acid rain?

_____

_____

_____

_____

4. Describe the extent of river pollution in Europe. Specifically, which European countries have polluted rivers?

_____

_____

_____

_____

Using Figure 8.31 "Industrial Regions of Europe" (p. 282) as a reference, use a red or black colored pencil (or another color that will contrast with the shading scheme you have created) to pattern in the older and newer European industrial areas (use a different pattern for each). Mark these symbols on the map's legend. Once you have done this, answer the following questions.

5. Describe the location of the older European industrial areas. Specifically, in which European countries can they be found?

_____

_____

_____

_____

6. What type of manufacturing constitutes "older" industrial activities?

_____

_____

7. Does there appear to be a geographic pattern associated with these older industrial areas? If so, describe the pattern.

_____

_____

_____

_____

8. Describe the location of the newer European industrial areas. Specifically, in which European countries can they be found?

_____

_____

_____

_____

9. What type of manufacturing constitutes "newer" industrial activities?

_____

_____

10. Does there appear to be a geographic pattern associated with these newer industrial areas? If so, describe the pattern.

_____

_____

_____

_____

11. Compare the location of the European industrial areas and the extent of acid rain in the region. Is there a correlation between the locations of these polluted regions and that of industrial activity?

_____

_____

_____

_____

12. Do the polluted areas tend to correlate more with that of the older or the newer industrial activities?

_____

_____

_____

_____

13. Explain this correlation.

_____

_____

_____

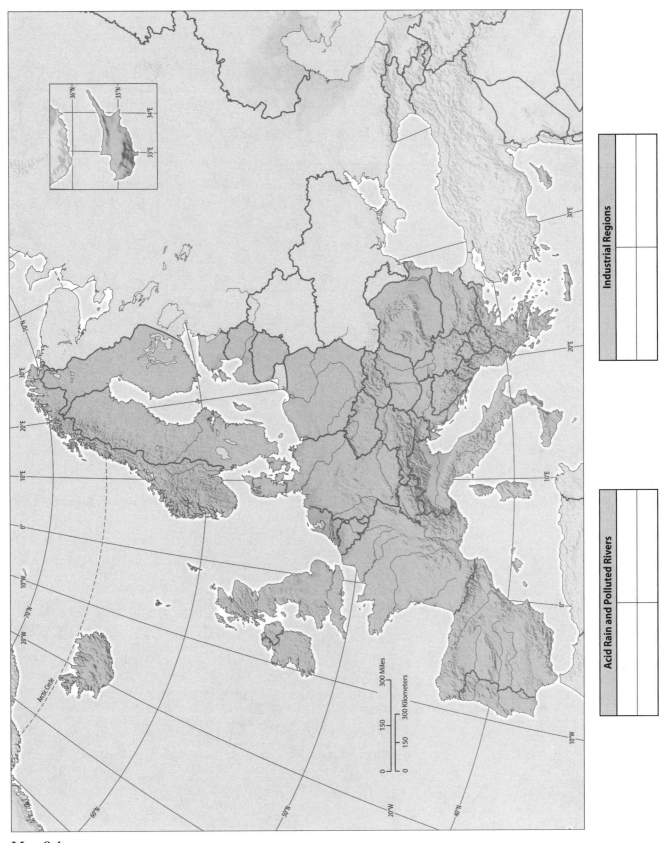

Industrial Regions

Acid Rain and Polluted Rivers

Map 8.4

Exercise Two: European Language and Religion Patterns

Using Figure 8.18 "Language Map of Europe" (p. 269), Figure 8.20 "Religions of Europe" (p. 271), and Mapping Workbook Map 8.5, complete the following exercise.

Using Figure 8.18 "Language Map of Europe" (p. 269) as a reference, use a colored pencil to shade in the region where the Germanic language subfamily is dominant. Using a different color for each, shade in the other major language subfamilies in their respective locations on the map. Mark the appropriate shading schemes in the map's legend. Once you have done this, answer the following questions.

1. Describe the locations of each of the language subfamilies in Europe. More specifically, do you notice any overall patterns of clustering of language subfamilies in Europe (e.g., are all the Germanic languages clustered together)?

   _____

   _____

   _____

   _____

   _____

   _____

2. If so, what do you think accounts for these patterns? Explain.

   _____

   _____

   _____

   _____

Using Figure 8.20 "Religions of Europe" (p. 271) as a reference, use a red or black (or another color that will contrast with the shading scheme you have created) colored pencil to pattern in the region where the majority of the population is Roman Catholic. Using a different color for each, pattern in the other major religions in their respective locations on the map. Mark the appropriate patterning scheme in the map's legend. Once you have done this, answer the following questions.

3. Describe the locations of each of the modern religions in Europe. More specifically, do you notice any overall patterns of clustering of religions in Europe (e.g., are all the Protestants clustered together)?

   _____

   _____

   _____

   _____

   _____

   _____

4. If so, what do you think accounts for these patterns? Explain.

_____

_____

_____

_____

5. Examine your mapped patterns of language and religion. Do you notice any correlations between certain language subfamilies and modern religions practiced in European countries? If so, in which countries do these correlations exist?

_____

_____

_____

_____

6. If you answered *yes* to question 5, what do you think accounts for the correlations?

_____

_____

_____

_____

7. Which European countries possess non-Indo-European language–speaking majorities?

_____

_____

8. Do you see a correlation between language and religion in the non-Indo-European language–speaking countries? If so, what are they and what do you think accounts for them? If not, why do you think they are different?

_____

_____

_____

_____

_____

Map 8.5

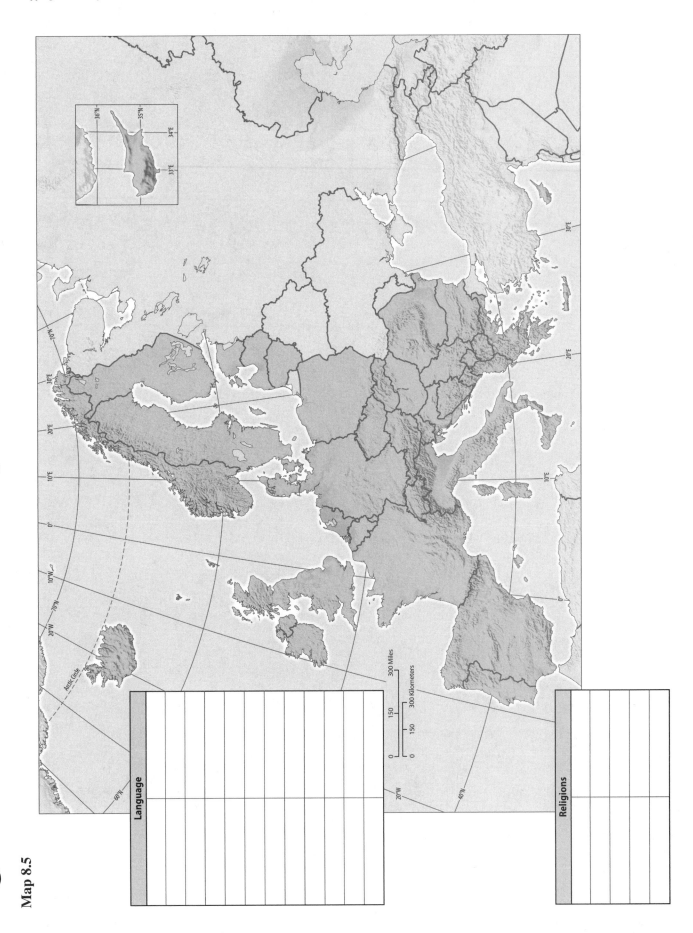

Language

Religions

300 Miles
300 Kilometers

Chapter Nine: The Russian Domain Mapping Workbook Exercises

Identify the following features on Mapping Workbook Maps 9.1, 9.2, and 9.3

Identify and label the following countries on Mapping Workbook Map 9.1

Belarus
Georgia
Moldova
Russia
Ukraine
Armenia

Identify and label the following cities on Mapping Workbook Map 9.2

Chelyabinsk	Kiev	St. Petersburg
Chernobyl	Krasnoyarsk	Tbilisi
Chisinau	Minsk	Verkhoyansk
Dnepropetrovsk	Moscow	Vladivostok
Donetsk	Murmansk	Volgograd
Groznyy	Norilsk	Yakutsk
Irkutsk	Novosibirsk	Yekaterinburg
Kaliningrad	Odessa	Yerevan
Kazan	Omsk	Nizhniy Novgorod
Khabarovsk	Petropavlovsk-Kamchatskiy	Novokuznetsk
Kharkov	Samara	

Identify and label the following physical features on Mapping Workbook Map 9.3

Amur River	Gulf of Finland	Sea of Okhotsk
Arctic Ocean	Kamchatka Peninsula	Siberia
Baltic Sea	Kola Peninsula	Ural Mountains
Barents Sea	Kuril Islands	Ussuri River
Bering Sea	Lake Baikal	Verkhoyansk Range
Black Sea	Lena River	Volga River
Caspian Sea	Novaya Zemlya (Island)	West Siberian Plain
Caucasus Mountains	Northeast Highlands	White Sea
Central Siberian Highland	Ob River	Yenisey River
Crimean Peninsula	Pacific Ocean	Yakutsk Basin
Dnieper River	Sakhalin Island	
European Plain	Sea of Azov	

Map 9.1

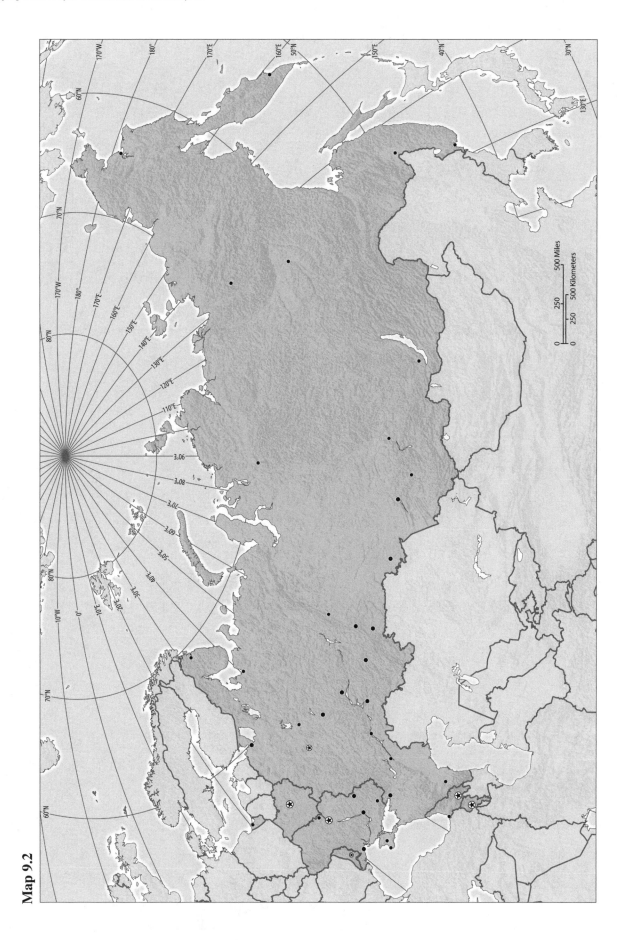

Map 9.2

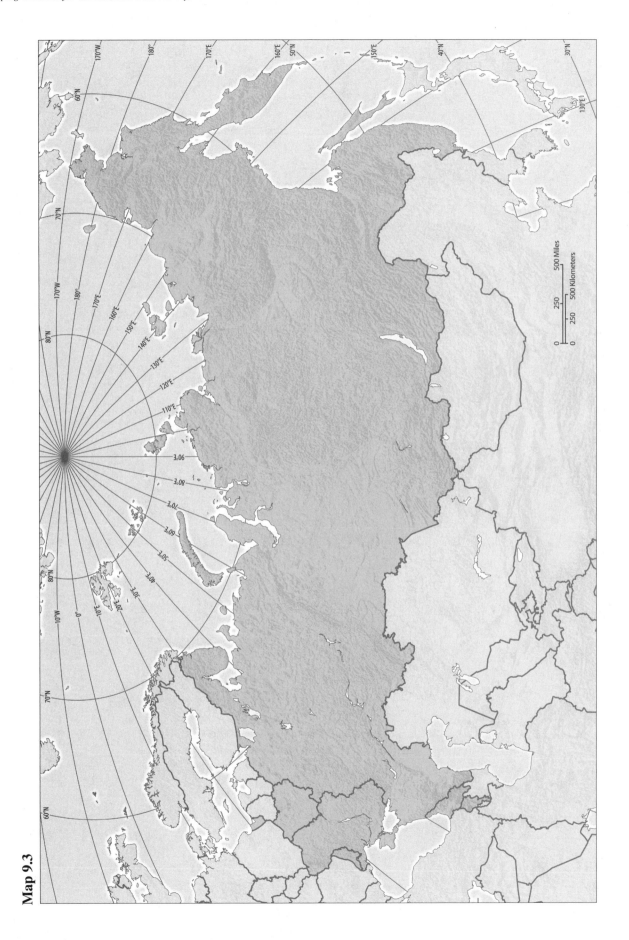

Map 9.3

Exercise One: Climate and Russian Domain Agricultural and Industrial Activities

Using Figure 9.7 "Climate Map of the Russian Domain" (p. 299), Figure 9.8 "Agricultural Regions" (p. 300), and Mapping Workbook Map 9.4, complete the following exercise.

Using Figure 9.7 "Climate Map of the Russian Domain" (p. 299) as a reference, use a colored pencil to shade in the regions that possess a Dfb climate. Using different colored pencils to represent each climate type, shade in the other major climates within the region. Mark this shading scheme in the map's legend. Once you have done this, answer the following questions.

1. What are the major climates of the Russian Domain?

_____

_____

2. What are the general locations of these climates?

_____

_____

Using Figure 9.8 "Agricultural Regions" (p. 300) as a reference, use a red or black (or another color that contrasts with the shading scheme you created for climate) colored pencil to pattern in the major agricultural types within the region, using different patterns for diversified agriculture, large-scale grain production, urban truck farming, and humid subtropical specialized agricultural production. Also, use a different pattern to indicate the locations of tundra, taiga, drylands, and mountains. Mark this patterning scheme in the map's legend. Once you have done this, answer the following questions.

3. What is (are) the location(s) of diversified agriculture within the Russian Domain?

_____

_____

4. What is (are) the location(s) of large-scale grain production within the Russian Domain?

_____

_____

5. What is (are) the location(s) of urban truck farming within the Russian Domain?

_____

_____

6. What is (are) the location(s) of humid subtropical specialized agriculture within the Russian Domain?

_____

_____

7. How well do these agricultural activities correlate with the biomes and landforms (i.e., tundra, taiga, drylands, and mountains) in the region?

_____

_____

_____

_____

8. Examine the agricultural regions and their location relative to the climate types, biomes, and landforms you shaded in on the map. Which biomes, landforms, and climate does diversified agriculture correlate with, if any?

_____

_____

9. Explain the correlation between this agricultural activity and the biomes, landforms, and climate within the region where it prevails? If there was no correlation, how would you explain the lack of one?

_____

_____

_____

_____

10. Which biomes, landforms, and climate correlate with large-scale grain production, if any?

_____

_____

11. How would you explain the correlation between this agricultural activity and the biomes, landforms, and climate within the region where it prevails? If there was no correlation, how would you explain the lack of one?

_____

_____

_____

_____

12. Which biomes, landforms, and climate correlate with urban truck farming, if any?

_____

_____

13. Explain the correlation between this agricultural activity and the biomes, landforms, and climate within the region where it prevails? If there was no correlation, how would you explain the lack of one?

_____

_____

_____

_____

14. Which biomes, landforms, and climate correlate with humid subtropical specialized agricultural production, if any?

_____

_____

15. How would you explain the correlation between this agricultural activity and the biomes, landforms, and climate within the region where it prevails? If there was no correlation, how would you explain the lack of one?

_____

_____

_____

_____

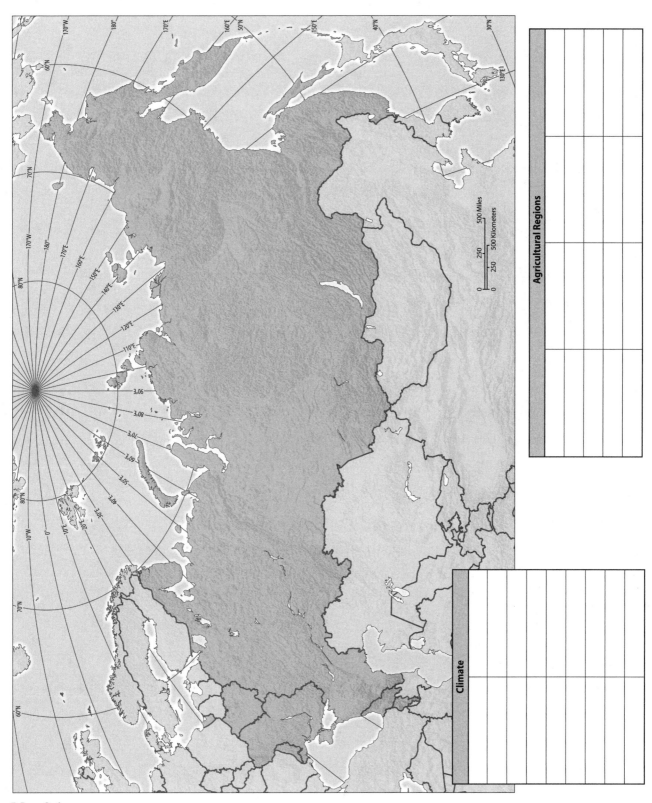

Agricultural Regions

Climate

Map 9.4

Exercise Two: Regional Patterns of Pollution and Industrialization in the Russian Domain

Using Figure 9.5 "Environmental Issues in the Russian Domain" (p. 298), Figure 9.35 "Major Natural Resources and Industrial Zones" (p. 319), and Mapping Workbook Map 9.5, complete the following exercise.

Using Figure 9.35 "Major Natural Resources and Industrial Zones" (p. 319) as a reference, use a colored pencil to shade in the major regions of forestry in the Russian Domain. Mark this shading scheme on the map's legend. Use a different colored pencil to shade in the major regions of manufacturing within the Russian Domain. Create symbols for the major natural resources found in the Russian Domain and label these on the map in their respective locations. Use a different colored pencil for each symbol. Mark these symbols on the map's legend. Once you have done this, answer the following questions.

1. Describe the location of the natural resources found in the Russian Domain. Specifically, in which regions are they found?

_____

_____

_____

_____

2. Where is the greatest concentration of natural resources in the Russian Domain?

_____

_____

Now examine the mapped location of the various industrial activities in the Russian Domain.

3. Do you see a correlation between the location of the Russian Domain's natural resources and the location of the Russian Domain's major industrial regions? If so, describe it. Explain this correlation.

_____

_____

_____

_____

4. Are there any areas where there are natural resources and yet there are few industrial activities? If so, where?

_____

_____

_____

_____

5. If you answered *yes* to question 4, what could explain this phenomenon?

_____

_____

_____

_____

_____

Using Figure 9.5 "Environmental Issues in the Russian Domain" (p. 298) as a reference, use a colored pencil to pattern in the regions of the Russian Domain that have been affected by acid rain. Mark this patterning scheme on the map's legend. Now, using a different colored pencil, create different patterns and indicate the regions of forest damage, radioactive contamination, coastal pollution, salinization, and polluted rivers. Mark these patterns in the map's legend. Once you have done this, answer the following questions.

6. Compare the locations of the Russian Domain industrial activities and that of the extent of acid rain in the region. Describe the correlation between the locations of these polluted regions and that of industrial activity.

_____

_____

7. Explain this correlation.

_____

_____

_____

_____

8. Compare the locations of the Russian Domain's industrial activities and that of the extent of forest damage in the region. Is there a correlation between the locations of these polluted regions and that of industrial activity?

_____

_____

9. If you found a correlation, describe and explain it.

_____

_____

_____

_____

10. Compare the locations of the Russian Domain's industrial activities and that of the extent of radioactive contamination in the region. Is there a correlation between the locations of these polluted regions and that of industrial activity?

_____

_____

11. If you found a correlation, describe and explain it.

_____

_____

_____

_____

12. Compare the locations of the Russian Domain's industrial activities and that of the extent of coastal pollution in the region. Is there a correlation between the locations of these polluted regions and that of industrial activity?

_____

_____

13. If you found a correlation, describe and explain it.

_____

_____

_____

_____

14. Compare the locations of the Russian Domain's industrial activities and that of the extent of salinization in the region. Is there a correlation between the locations of these polluted regions and that of industrial activity?

_____

_____

15. If you found a correlation, describe and explain it.

_____

_____

_____

_____

16. Compare the locations of the Russian Domain's industrial activities and that of the extent of river pollution in the region. Is there a correlation between the locations of these polluted regions and that of industrial activity?

_____

_____

17. If you found a correlation, describe and explain it.

_____

_____

_____

_____

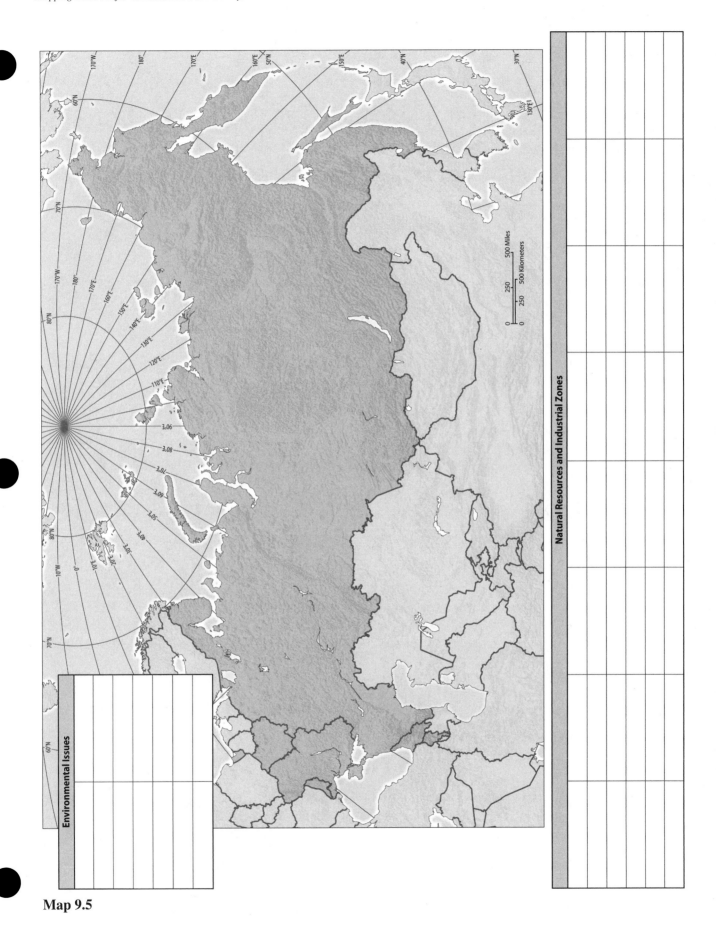

Natural Resources and Industrial Zones

Environmental Issues

Map 9.5

Chapter Ten: Central Asia Mapping Workbook Exercises

Identify the following features on Mapping Workbook Maps 10.1, 10.2, and 10.3

Identify and label the following countries on Mapping Workbook Map 10.1

Afghanistan
Azerbaijan
Kazakhstan
Kyrgyzstan
Mongolia
Tajikistan
Turkmenistan
Uzbekistan

Identify and label the following cities on Mapping Workbook Map 10.2

Almaty	Bishkek	Nukus
Aqtau	Dushanbe	Oral
Ashgabat	Hohhot	Shymkent
Astana	Kabul	Tashkent
Baku	Kandahar	Ulaanbaatar
Baotou	Lhasa	Urumqi

Identify and label the following physical features on Mapping Workbook Map 10.3

Altai Mountains	Kara-Bogaz Gol	Syr Darya River
Altan Shan	Karakoram Range	Taklamakan Desert
Amu Darya River	Kara Kum Canal	Tarim Basin
Aral Sea	Kara Kum Desert	Tarim River
Caspian Sea	Kunlun Shan	Tibetan Plateau
Gobi Desert	Kura River	Tien Shan
Helmand River	Kyzyl Kum Desert	Turfan Depression
Hindu Kush	Lake Balqash	Ural River
Huang He River	Pamir Knot	
Ili River	Pamir Mountains	

Map 10.1

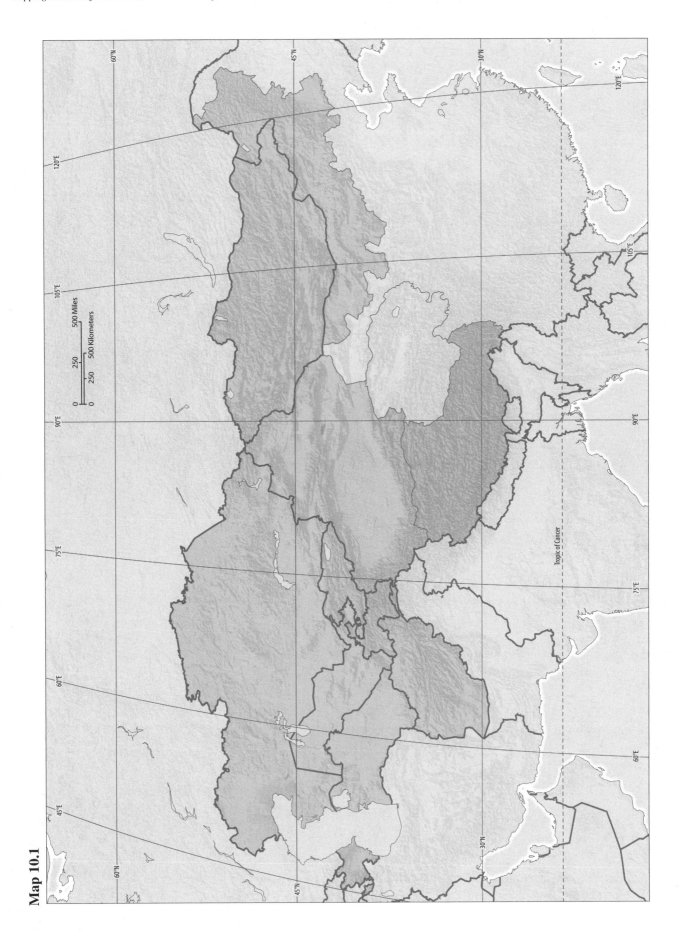

Map 10.2

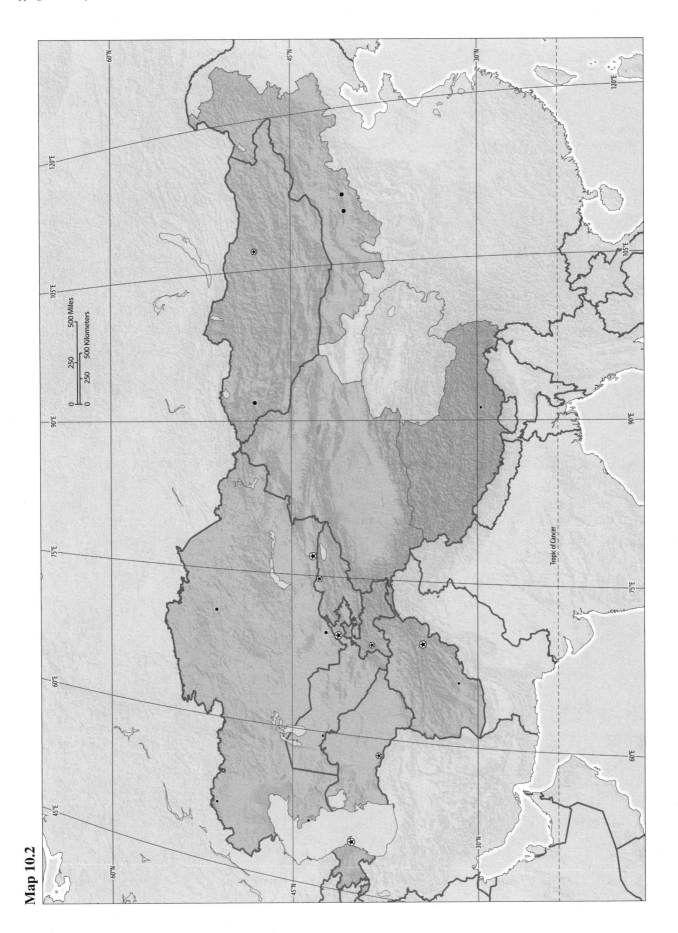

Map 10.3

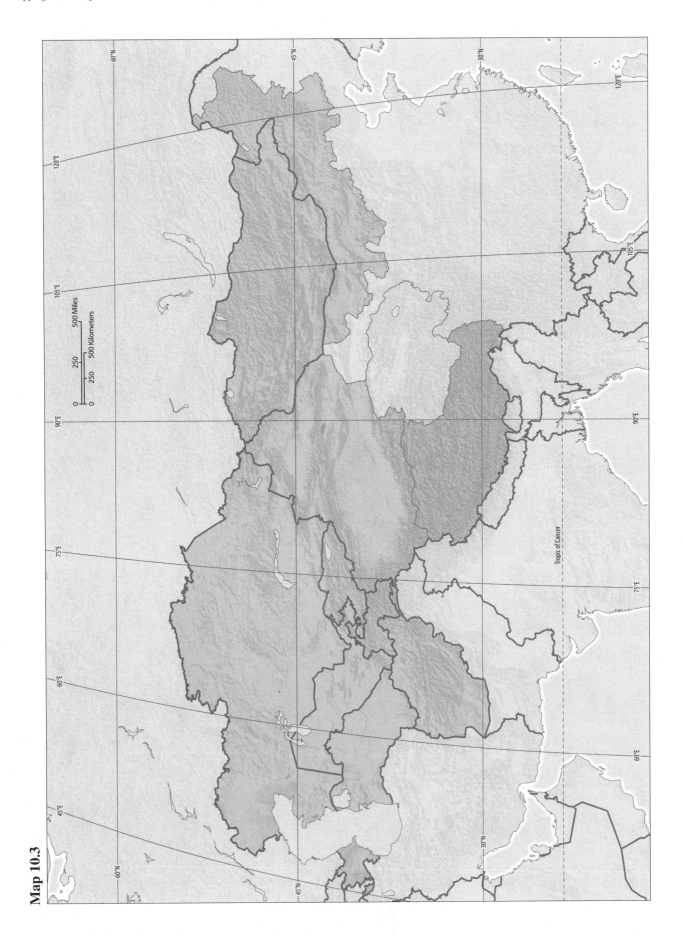

Exercise One: Landforms, Climate, and Settlement in Central Asia

Using Figure 10.1 "Central Asia" (physical features) (pp. 328–329), Figure 10.6 "Climates of Central Asia" (p. 334), Figure 10.8 "Population Density in Central Asia" (p. 335), Figure 10.9 "Population Patterns in Xinjiang's Tarim Basin" (p. 336), and Mapping Workbook Map 10.4, complete the following exercise.

Using Figure 10.1 "Central Asia" (physical features) (pp. 328–329) as a reference, use a colored pencil to shade in the major Central Asian desert regions. Using different colored pencils to represent each landform, shade in the Central Asian highlands, steppes, and plateaus in their appropriate locations. Use a black or red (or another color that will contrast with the shading scheme you have created) colored pencil and draw lines indicating the approximate location of the rivers in the region. Mark the shades and symbols for the landforms and rivers in the map's legend. Once you have done this, answer the following questions.

1. Describe the location of the deserts in Central Asia.

_____

_____

_____

_____

2. What general criteria are used to classify a region as a desert?

_____

_____

3. What is a steppe?

_____

_____

4. Describe the location of the steppes in Central Asia.

_____

_____

_____

_____

5. Describe the location of the highlands in Central Asia.

_____

_____

_____

_____

6. How do the locations of rivers generally correspond to the locations of highlands in the region?

_____

_____

Using Figure 10.6 "Climates of Central Asia" (p. 334) as a reference, use colored pencils to pattern in their respective locations the major Central Asian climate types. Use a different color to represent each climate type. Mark the patterns for the climates in the map's legend. Once you have done this, answer the following questions.

7. Describe the general climatic pattern in Central Asia. What is the predominant climate in the region?

_____

_____

_____

_____

_____

8. Examine the pattern of Central Asian climate and the pattern of Central Asian landforms together. Are there correlations between the location of the various landforms and the major climates in the region? If so, describe what they are.

_____

_____

_____

_____

_____

_____

9. How would the landforms within Central Asia have an effect (i.e., control) on the dominant climate in the region?

_____

_____

_____

_____

10. How might the landforms also cause more local variations in the Central Asian climate patterns?

_____

_____

_____

_____

Using Figure 10.8 "Population Density in Central Asia" (p. 335) as a reference, use a red or black (or another color that contrasts with the shading and patterning schemes you have already established) colored pencil and color in the Central Asian regions possessing high concentrations of population. Mark the shades or pattern for population concentration in the map's legend. Once you have done this, answer the following questions.

11. Describe the regions of highest population concentration.

_____

_____

12. Describe the regions of lowest population concentration.

_____

_____

Examine the regions of the highest and lowest population concentration and how this concentration corresponds to the locations of the various landforms and climates you placed on your map.

13. Describe how Central Asian population correlates with landforms. Explain this correlation.

_____

_____

_____

_____

14. Describe how Central Asian population correlates with climate. Explain this correlation.

_____

_____

_____

_____

Look at Figure 10.9 "Population Patterns in Xinjiang's Tarim Basin" (p. 336), closely paying attention to the pattern of population settlement and landforms.

15. Does the settlement within the Tarim Basin reflect the wider patterns of Central Asian population settlement?

_____

_____

16. If so, explain why you believe this is the case. If not, explain why you think population settlement within the Tarim Basin differs from the rest of the region.

_____

_____

_____

_____

_____

_____

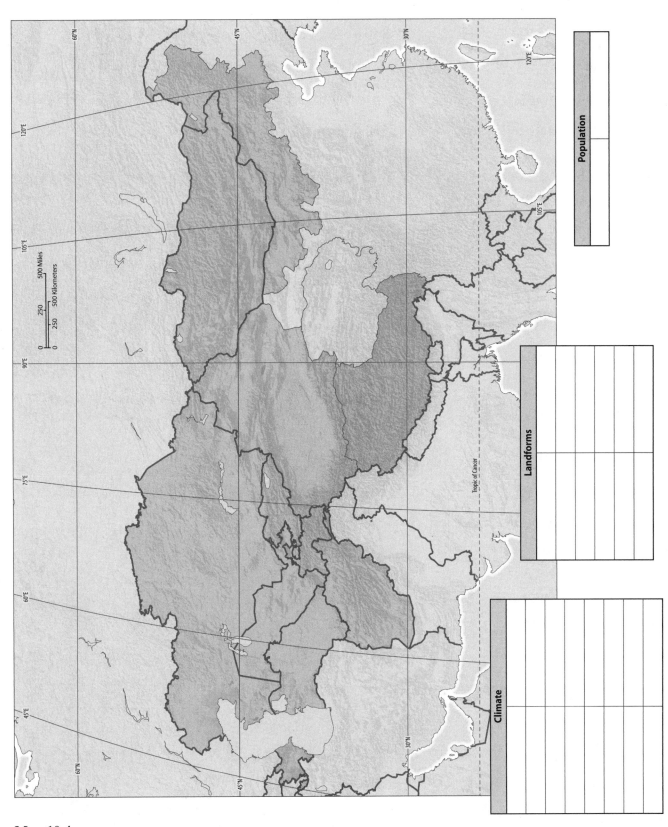

Map 10.4

Exercise Two: Valley-ism in Afghanistan

Using Figure 10.1 "Central Asia" (physical features) (pp. 328–329), Figure 10.16 "Afghanistan's Ethnic Patchwork" (p. 340), Figure 10.8 "Population Density in Central Asia" (p. 335), and Mapping Workbook Map 10.5, complete the following exercise.

Using Figure 10.16 "Afghanistan's Ethnic Patchwork" (p. 340) as a reference, use a colored pencil to shade in the regions that possess the majority of the Tajik speakers. Using a different color to represent each language, indicate the location of these other languages on your map by shading them in their respective regions. Mark the shades in the map's legend. Once you have done this, answer the following.

Describe the general patterns of each of the major languages within Central Asia.

1. Tajik _____

_____

2. Hazara _____

_____

3. Aimak _____

_____

4. Pashtun _____

_____

5. Nuristani _____

_____

6. Pashai _____

_____

7. Baluchi _____

_____

8. Uzbek _____

_____

9. Turkmen _____

_____

10. Kyrgyz _____

_____

11. Brahui _____

_____

12. Which of these languages displays the greatest amount of overlap?

_____

_____

13. In the regions where this overlap occurs, how different are these languages from each other?

_____

_____

14. How might the linguistic differences cause cultural and political tensions in these areas?

_____

_____

Using Figure 10.1 "Central Asia" (physical features) (pp. 328–329) as a reference, use a red or black (or another color that would contrast with the shading scheme you have established) colored pencil to pattern in the major Afghanistan landforms. Mark the patterns in the map's legend. Once you have done this, answer the following questions.

15. What is the predominant landform in Afghanistan?

_____

_____

16. How extensive is this landform in Afghanistan?

_____

_____

17. Describe the pattern of mountains in Afghanistan. Which way do the mountains trend (e.g., north–south, east–west, etc.)?

_____

_____

_____

Using Figure 10.8 "Population Density in Central Asia" (p. 335) as a reference, use a red or black (or another color that would contrast with the shading scheme you have established) colored pencil to pattern in the Afghanistan population distribution. Mark the pattern in the map's legend. Once you have done this, answer the following questions.

18. Examine the physical geography of Afghanistan and compare it with the pattern of population settlement in the country. How does settlement correlate with Afghanistan's physical geography?

_____

_____

19. How do you explain this correlation?

_____

_____

20. Examine the physical regions, the patterns of population settlement, and patterns of language you drew on your map. How do the patterns of population settlement correlate with the patterns of language in the country?

_____

_____

21. How do you believe these patterns affect the cultural relations, political cohesion, and stability within the country?

_____

_____

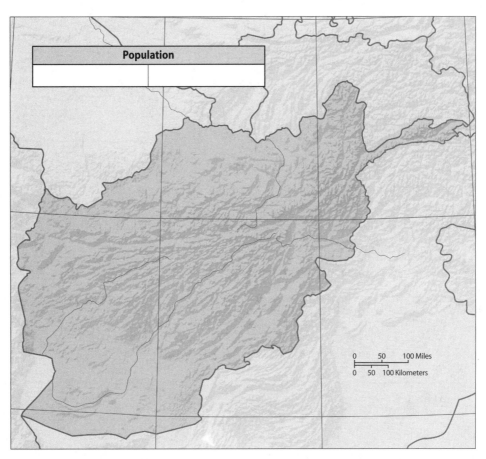

Population	

Languages	

Landforms	

Map 10.5

Chapter Eleven: East Asia Mapping Workbook Exercises

Identify the following features on Mapping Workbook Maps 11.1, 11.2, and 11.3

Identify and label the following countries on Mapping Workbook Map 11.1

China
Japan
North Korea
South Korea
Taiwan

Identify and label the following cities on Mapping Workbook Map 11.2

Beijing	Hong Kong	Shenyang
Busan	Kyoto	Shenzhen
Changsha	Lhasa	Taipei
Chengdu	Macao	Tangshan
Chongqing	Nanjing	Tianjin
Guangju	Osaka-Kobe	Tokyo
Guangzhou	Pyongyang	Urumqi
Hangzhou	Sapporo	Wuhan
Haerbin	Seoul	Xi'an
Hiroshima	Shanghai	

Identify and label the following physical features on Mapping Workbook Map 11.3

Amur River	Hong (Red) River	Plateau of Tibet
Bay of Bengal	Honshu	Ryukyu Islands
East China Sea	Huang He	Sea of Japan
Gobi Desert	Jeju (Island)	Shikoku
Grand Canal	Kyushu	South China Sea
Gulf of Tonkin	Loess Plateau	Taklamakan Desert
Hainan (Island)	Mekong River	Ussuri River
Himalayas	Okinawa	Yangtze River
Hokkaido	Pacific Ocean	Yellow Sea

Map 11.1

Map 11.2

Map 11.3

Exercise One: Populated and Unpopulated Regions of China

Using Figure 11.2 "East Asia" (physical landscape) (p. 357), Figure 11.3 "Environmental Issues in East Asia" (p. 359), Figure 11.16 "Population Map of East Asia" (p. 367), and Mapping Workbook Map 11.4, complete the following exercise.

Using Figure 11.2 "East Asia" (physical landscape) (p. 357) as a reference, use a colored pencil to shade in the low-elevation regions of China (0–499 feet above sea level). Using a different colored pencil, shade in the higher-elevation regions of China (500–1,999 feet above sea level). Using a different colored pencil, shade in the higher-elevation regions of China (2,000–3,999 feet above sea level). Using a pencil that is a different color from the first two you used, shade in the highest elevation regions of China (4,000+ feet above sea level). Mark the shading scheme in the map's legend. Once you have done this, answer the following questions.

1. Describe the location of the low-elevation regions in China.

   _____

   _____

   _____

2. Describe the location of the higher-elevation regions in China.

   _____

   _____

   _____

3. Describe the location of the highest elevation regions in China.

   _____

   _____

   _____

Using Figure 11.16 "Population Map of East Asia" (p. 367) as a reference, use a red or black (or a color that will contrast with the elevation shading pattern you established) colored pencil to pattern the locations of Chinese population. Mark the patterning scheme in the map's legend. Once you have done this, answer the following questions.

4. What regions in China possess the highest population densities?

   _____

   _____

5. What do you think accounts for the clustering of people in these regions?

   _____

   _____

6. Compare the patterns of elevation and population settlement on your map. Do physical features seem to play a role in high population densities? If so, what are the physical characteristics of the landscape that appear to promote population?

   _____

   _____

7. Where are the Asian regions that possess the lowest population densities?

   _____

   _____

8. What do you think accounts for the clustering of people in these regions?

   _____

   _____

9. Do physical features seem to play a role in low population densities? If so, what are the physical characteristics of the landscape that appear to limit population?

_____

_____

Using Figure 11.3 "Environmental Issues in East Asia" (p. 359) as a reference, use a colored pencil to pattern the location of forest areas in China. Using a different colored pencil for each, pattern in the Chinese regions of extensive deforestation, desertification, severe soil erosion, coastal pollution, and high risk of flooding. Mark the patterning scheme in the map's legend. Once you have done this, answer the following questions.

10. Where are the greatest locations of extensive deforestation in China?

_____

_____

11. Where are the greatest locations of desertification in China?

_____

_____

12. Where are the greatest locations of severe soil erosion in China?

_____

_____

13. Where are the greatest locations of coastal pollution in China?

_____

_____

14. Examine the location of the above environmental hazards and how they correspond to population within China. Does there appear to be a correlation between the regions of highest population density and the location of these environmental hazards? If so, specifically describe these correlations.

_____

_____

15. How are these environmental impacts related to population? Elaborate.

_____

_____

16. Examine the location of the regions of China subject to the greatest risk of flooding and how it corresponds to population within China. Where are the regions of greatest flooding risk?

_____

_____

17. Does there appear to be a correlation between the regions of highest population density and the location of flooding risk? If so, specifically describe this correlation.

_____

_____

18. In what ways will flooding have an impact on the Chinese population?

_____

_____

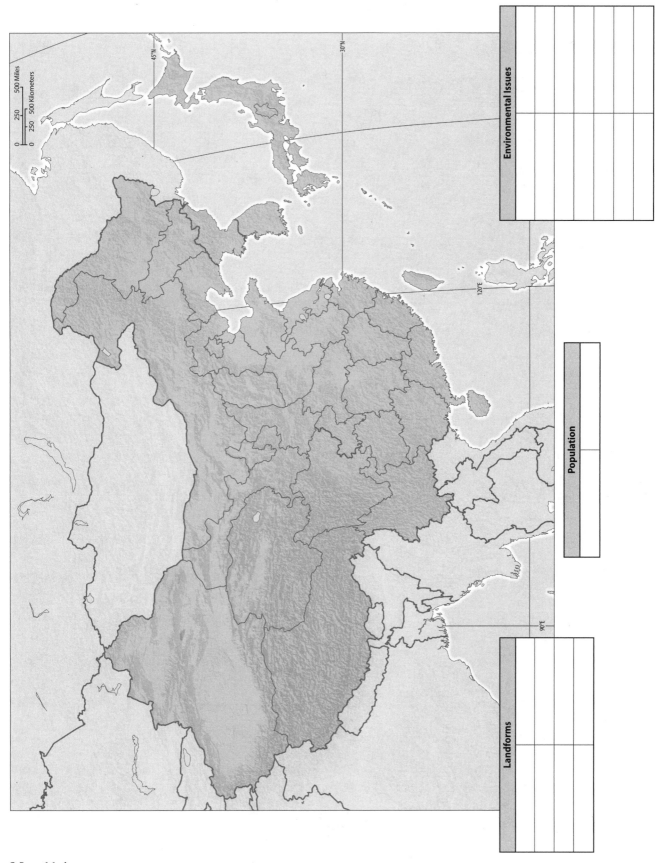

Map 11.4

Exercise Two: Japanese Physical Geography and Settlement

Using Figure 11.9 "Japan's Physical Geography" (p. 363), Figure 11.16 "Population Map of East Asia" (p. 367), Figure 11.20 "Urban Concentration in Japan" (p. 369), and Mapping Workbook Map 11.5, complete the following exercise.

Using Figure 11.9 "Japan's Physical Geography" (p. 363) as a reference, use a colored pencil to shade in the approximate locations of hill lands and mountains in Japan. Using a different colored pencil shade in the approximate locations of diluvial plains and lowlands of new alluvium in Japan. Using a different colored pencil, draw lines representing the approximate locations of plate boundaries. Label the plates and plate boundaries. Using a different colored pencil, draw lines indicating the regions of tsunami activity within Japan. Using different colored pencils, create point symbols and indicate the location for earthquake epicenters and major volcanic eruptions. Mark the shading scheme and symbols on the map's legend. Once you have done this, answer the following questions.

1. What are the names of the tectonic plates that converge on Japan?

   _____

   _____

2. Does the location of these plates correspond to the location of volcanic and earthquake activity? If so, what may account for it?

   _____

   _____

   _____

   _____

Using Figure 11.16 "Population Map of East Asia" (p. 367) as a reference, use a red or black (or a different color that will contrast with the shading scheme you created) colored pencil to indicate the Japanese regions of highest population density. Mark the population shading scheme on the map's legend. Once you have done this, answer the following questions.

3. Where are the areas of highest population density in Japan?

   _____

   _____

4. Where are the areas of lowest population density in Japan?

   _____

   _____

5. Examine the patterns of population and physical features you shaded on your map. Do you notice a correlation between the location of physical features and population concentration? If so, what are they?

   _____

   _____

   _____

   _____

6. What accounts for the patterns of population settlement in Japan?

   _____

   _____

   _____

Using Figure 11.20 "Urban Concentration in Japan" (p. 369) as a reference, use a red or black (or a different color that will contrast with the shading scheme you created) colored pencil to indicate the Japanese region's urban concentration. Mark the population shading scheme on the map's legend. Once you have done this, answer the following questions.

7. How does the location of this major urbanized Japanese region correlate to the physical features you labeled on your map?

_____

_____

_____

_____

8. How has the physical geography influenced the location and concentration of Japanese urban settlement? Explain.

_____

_____

_____

_____

9. How does the pattern of Japanese urban concentration correlate with the locations of earthquake epicenters?

_____

_____

_____

_____

10. How does the pattern of Japanese urban concentration correlate with the locations of major volcanic eruptions?

_____

_____

_____

_____

11. How does the pattern of Japanese urban concentration correlate with the locations of greatest potential tsunami activity?

_____

_____

_____

_____

12. How might any of these hazards (i.e., earthquakes, volcanic eruptions, and tsunamis) affect the Japanese population?

_____

_____

_____

_____

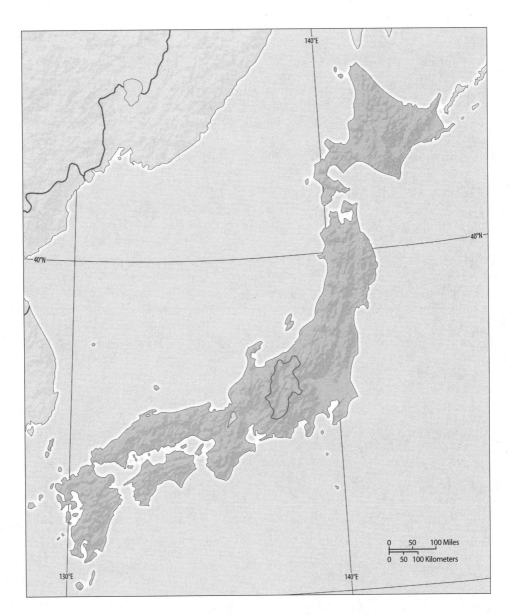

Landforms	

Population	

Urban Concentration	

Map 11.5

Chapter Twelve: South Asia Mapping Workbook Exercises

Identify the following features on Mapping Workbook Maps 12.1, 12.2, and 12.3

Identify and label the following countries on Mapping Workbook Map 12.1

Bangladesh
Bhutan
India
Maldives
Nepal
Pakistan
Sri Lanka

Identify and label the following cities on Mapping Workbook Map 12.2

Agra	Gwadar	Kolkata (Calcutta)
Ahmedabad	Hyderabad (India)	Lahore
Amritsar	Hyderabad (Pakistan)	Male
Bengaluru (Bangalore)	Islamabad	Multan
Chennai (Madras)	Jaffna	Mumbai (Bombay)
Chittagong	Jaipur	Nagpur
Colombo	Kanpur	New Delhi
Delhi	Karachi	Surat
Dhaka	Kathmandu	Thimphu

Identify and label the following physical features on Mapping Workbook Map 12.3

Andaman Islands	Ganges River	Narmada River
Andaman Sea	Godavari River	Nicobar Islands
Arabian Sea	Godwin Austen Peak (K2)	Palk Strait
Aravalli Range	Gulf of Khambhat	Rann of Kutch
Bay of Bengal	Gulf of Kutch	Ravi River
Bhima River	Indian Ocean	Satpura Range
Brahmaputra River	Indus River	Sulaiman Range
Cape Comorin	Jhelum River	Sundarbans
Central Makran Range	Karakoram Range	Sutlej River
Coromandel Coast	Kashmir	Thar Desert
Deccan Plateau	Kathiawar Peninsula	Vindhya Range
Eastern Ghats	Krishna River	Western Ghats
Gaghara River	Lakshadweep	Yamuna River
Ganges Delta	Malabar Coast	
Ganges Plain	Mt. Everest	

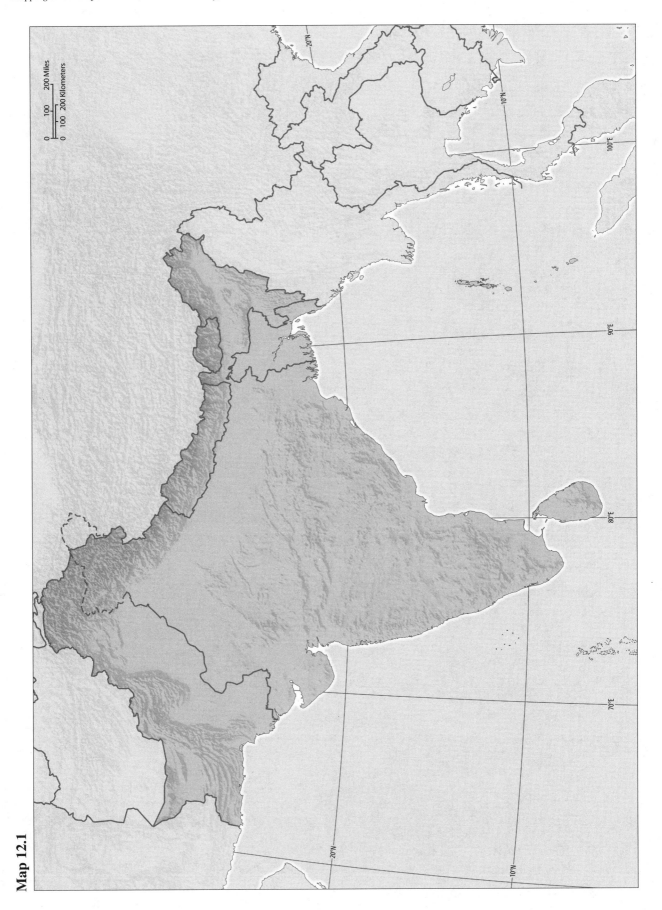

Map 12.1

Map 12.2

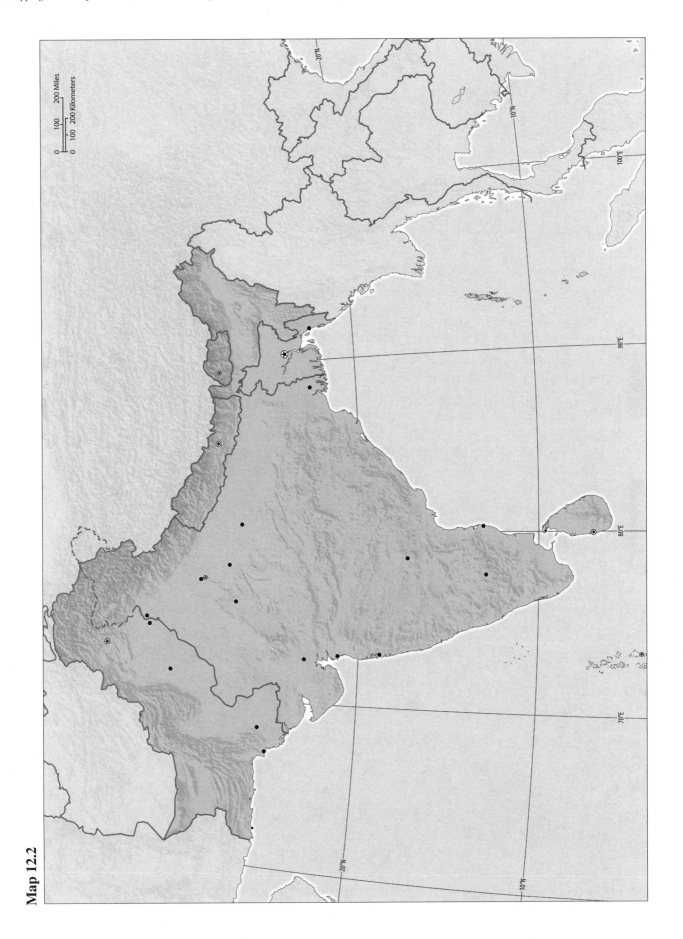

Map 12.3

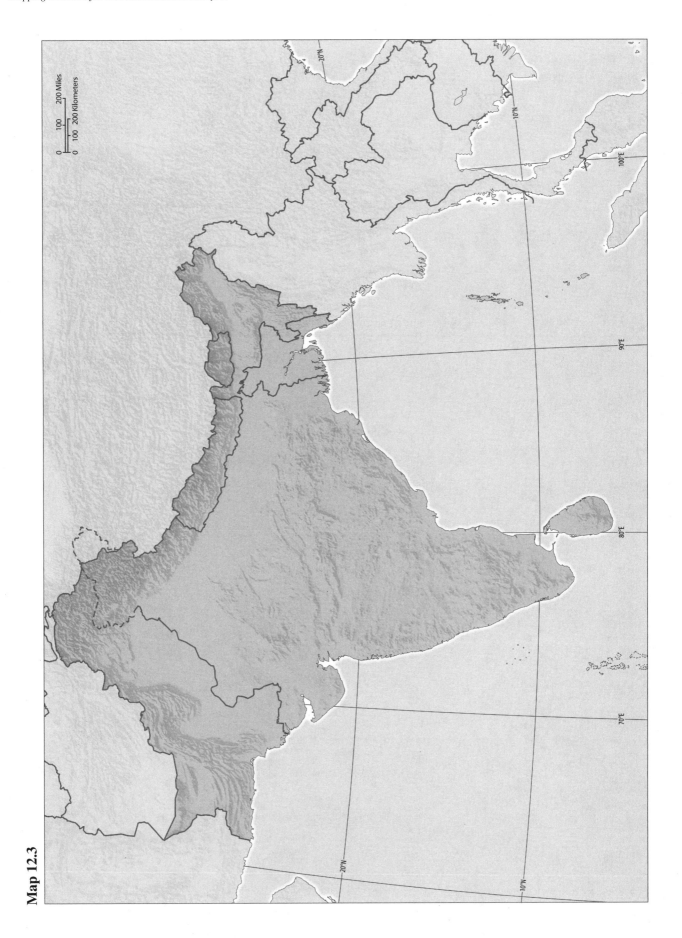

Exercise One: South Asian Patterns of Religion and Territorial Tension

Using Figure 12.19 Religious Geography of South Asia (p. 406), Figure 12.27 Geopolitical Issues in South Asia (p. 411), Figure 12.28 Geopolitical Change (p. 412), and Mapping Workbook Map 12.4, complete the following exercise.

Using Figure 12.19 "Religious Geography of South Asia" (p. 406) as a reference, use a colored pencil to shade in the major regions of Hinduism in South Asia. Using a different colored pencil for each religion, indicate the regions where the other major religions are prevalent in South Asia. Mark your shading scheme in the map's legend. Once you have done this, answer the following questions.

1. In which South Asian region(s) do you find the majority of Hindus?

_____

_____

2. In which South Asian region(s) do you find the majority of Muslims?

_____

_____

3. In which South Asian region(s) do you find the majority of Buddhists?

_____

_____

4. In which South Asian region(s) do you find the majority of Sikhs?

_____

_____

5. In which South Asian region(s) do you find the majority of Christians?

_____

_____

6. In which South Asian region(s) do you find the majority of Jains?

_____

_____

7. In which South Asian region(s) do you find the majority of adherents to tribal religions?

_____

_____

8. Do the adherents to these religions appear to be clustered within their own groups, or are they rather more evenly mixed throughout the region? Elaborate.

_____

_____

_____

_____

Using Figure 12.27 "Geopolitical Issues in South Asia" (p. 411) as a reference, use a colored pencil to pattern in the areas claimed by India, controlled by China. Using a different colored pencil and pattern to represent each conflict, indicate the locations of these conflicts on the South Asia map. Mark your patterning scheme in the map's legend. Once you have done this, answer the following questions.

9. Describe the various territorial conflicts in South Asia. Be sure to note their locations.

_____

_____

_____

_____

10. On the map you created, examine the pattern of religions and territorial conflict together. Do you see a correlation between any of the locations of the major South Asian religions and the locations of territorial conflict?

_____

_____

11. If so, what and where are they? Describe the nature of the conflicts.

_____

_____

_____

_____

_____

In your textbook, examine Figure 12.28 "Geopolitical Change" (p. 412) and study the evolution of the modern South Asian national borders.

12. How might the territorial progress of South Asia from a unified empire in 1700 to the collection of countries that it possesses today have generated the contemporary pattern of territorial conflicts? Elaborate on the location of these conflicts.

_____

_____

_____

_____

_____

_____

_____

13. Are these conflicts merely a matter of territory and shared borders, or does religion play a role? Elaborate.

_____

_____

_____

_____

_____

Map 12.4

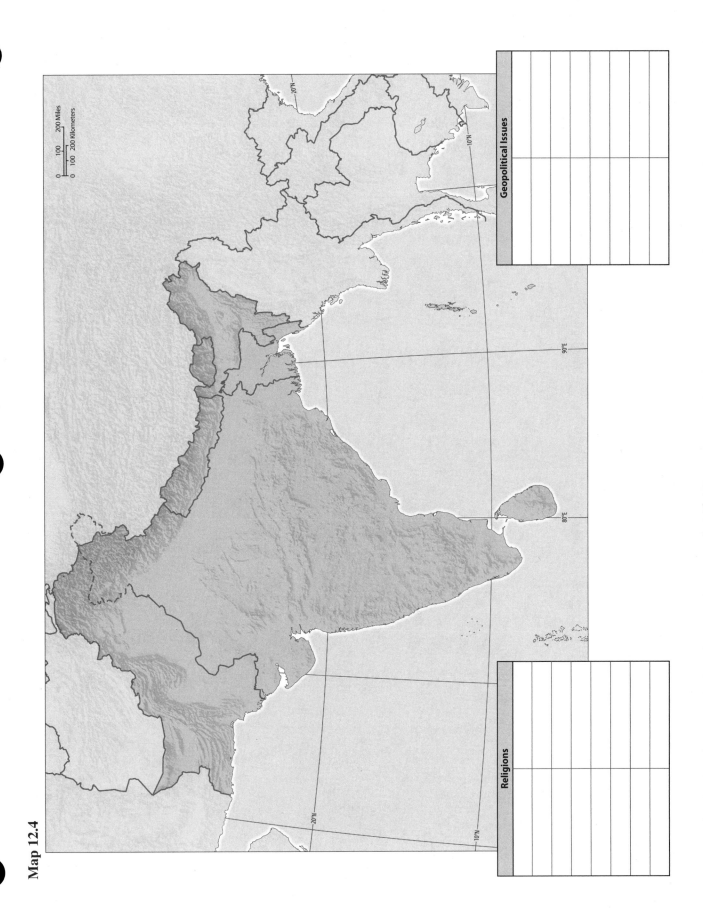

Religions

Geopolitical Issues

Exercise Two: South Asian Physical Geography and Population

Using Figure 12.2 "South Asia" (physical landscape) (p. 393), Figure 12.9 "Population Map of South Asia" (p. 400), and Mapping Workbook Map 12.5, complete the following exercise.

Using Figure 12.2 "South Asia" (physical landscape) (p. 393) as a reference, use a colored pencil to shade in the low-elevation regions of South Asia (0–499 feet above sea level). Using a different colored pencil, shade in the higher-elevation regions of South Asia (500–1,999 feet above sea level). Using a different colored pencil, shade in the higher-elevation regions of South Asia (2,000–3,999 feet above sea level). Using a pencil that is a different color from the first two you used, shade in the highest elevation regions of South Asia (4,000+ feet above sea level). Mark the shading scheme in the map's legend. Now, label the major mountain ranges, plains, plateaus, and deserts. Use a different colored pencil to draw in and label the location of the major South Asian rivers. Mark the symbol for rivers in the map's legend. Once you have done this, answer the following questions.

1. Describe the location of the low-elevation regions in South Asia.

   _____

   _____

   _____

   _____

2. Describe the location of the higher-elevation regions in South Asia.

   _____

   _____

   _____

   _____

3. Describe the location of the highest elevation regions in South Asia.

   _____

   _____

   _____

   _____

4. How does the location of rivers correlate with the high and low elevations within the region?

   _____

   _____

   _____

   _____

Using Figure 12.9 "Population Map of South Asia" (p. 400) as a reference, use a red or black (or a color that will contrast with the elevation shading pattern you established) colored pencil to pattern the locations of

South Asian population settlement. Mark the patterning scheme in the map's legend. Once you have done this, answer the following questions.

5. Where are the South Asian regions that possess the highest population densities?

_____

_____

6. What do you think accounts for the clustering of people in these regions?

_____

_____

7. Compare the patterns of elevation and population settlement on your map. Do physical features seem to play a role in high population densities? If so, what are the physical characteristics of the landscape that appear to promote population?

_____

_____

_____

_____

8. Where are the Asian regions that possess the lowest population densities?

_____

_____

9. What do you think accounts for the clustering of people in these regions?

_____

_____

10. Do physical features seem to play a role in low population densities? If so, what are the physical characteristics of the landscape that appear to limit population settlement?

_____

_____

_____

_____

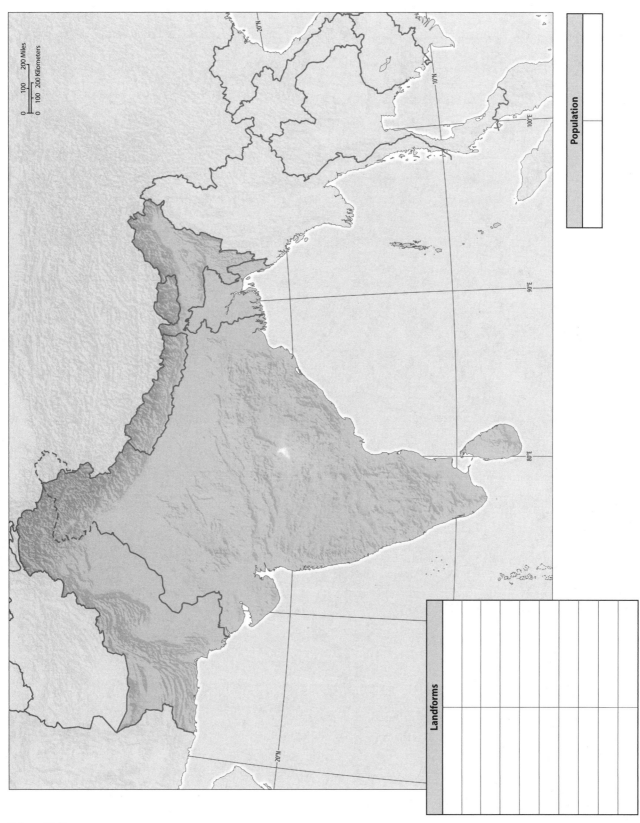

Population

Landforms

Map 12.5

Chapter Thirteen: Southeast Asia Mapping Workbook Exercises

Identify the following features on Mapping Workbook Maps 13.1, 13.2, and 13.3

Identify and label the following countries on Mapping Workbook Map 13.1

Brunei	Malaysia	Singapore
Burma (Myanmar)	Papua New Guinea	Thailand
Cambodia	Philippines	Timor-Leste
Indonesia	Sabah (state)	Vietnam
Laos	Sarawak (state)	

Identify and label the following cities on Mapping Workbook Map 13.2

Ambon	Ho Chi Minh City	Phnom Penh
Bandar Seri Begawan	Ipoh	Quezon City
Bandung	Jakarta	Semarang
Bangkok	Kuala Lumpur	Singapore
Banjarmasin	Kuching	Songkhia
Cebu	Makassar	Sorong
Da Nang	Mandalay	Surabaya
Davao	Manila	Vientiane
Dili	Medan	Yangon
Haiphong	Naypyidaw	
Hanoi	Palembang	

Identify and label the following physical features on Mapping Workbook Map 13.3

Rivers and Other Bodies of Water

Andaman Sea	Irrawaddy River	Red River
Arakan Coast	Indian Ocean	Salween River
Bay of Bengal	Java Sea	South China Sea
Celebes Sea	Mekong Delta	Strait of Malacca
Chao Phraya River Banda Sea	Mekong River	Sunda Strait
Gulf of Thailand	Pacific Ocean	Tonle Sap
Gulf of Tonkin	Philippine Sea	
Irrawaddy Delta	Red Delta	

Islands and Island Chains

Ambon	Madura	Sulawesi
Bali	Mindanao	Sumatra
Borneo	Negros	Sumba
Flores	Palawan	Sumbawa
Java	Papua	Timor
Lesser Sunda Islands	Paracel Islands	Visayas
Lombok	Seram	
Luzon	Spratly Islands	

Mountains, Mountain Ranges, and Miscellaneous Physical Features

Annam Mountains	Iran Mountains	Mt. Mayon
Arakan Mountains	Khorat Plateau	Mt. Pinatubo
Barisan Mountains	Mt. Kerinci	Mt. Semeru
Dieng Plateau	Mt. Kinabalu	Mt. Victoria

Map 13.1

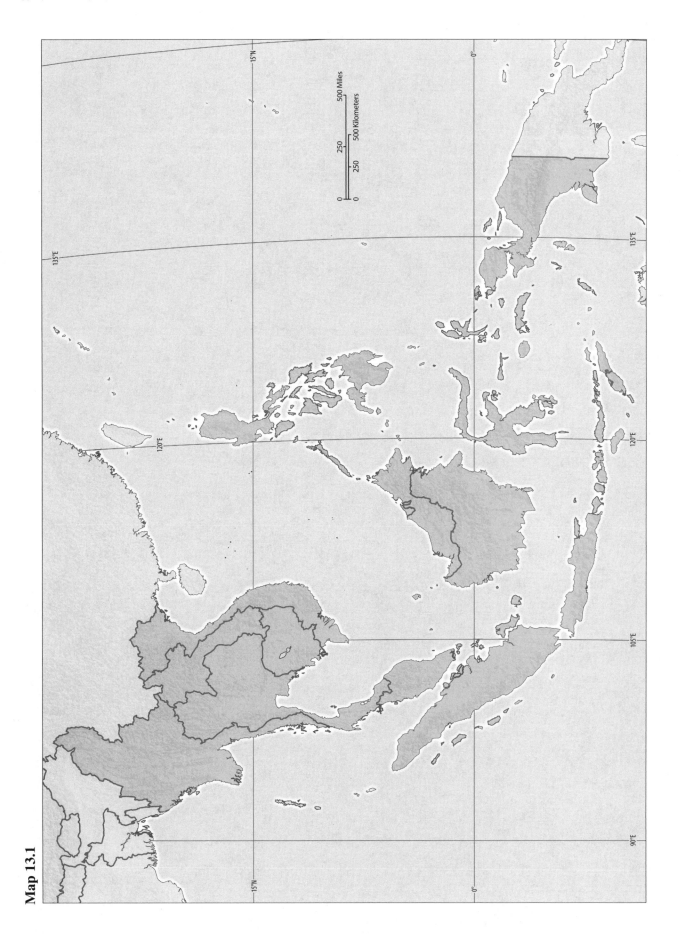

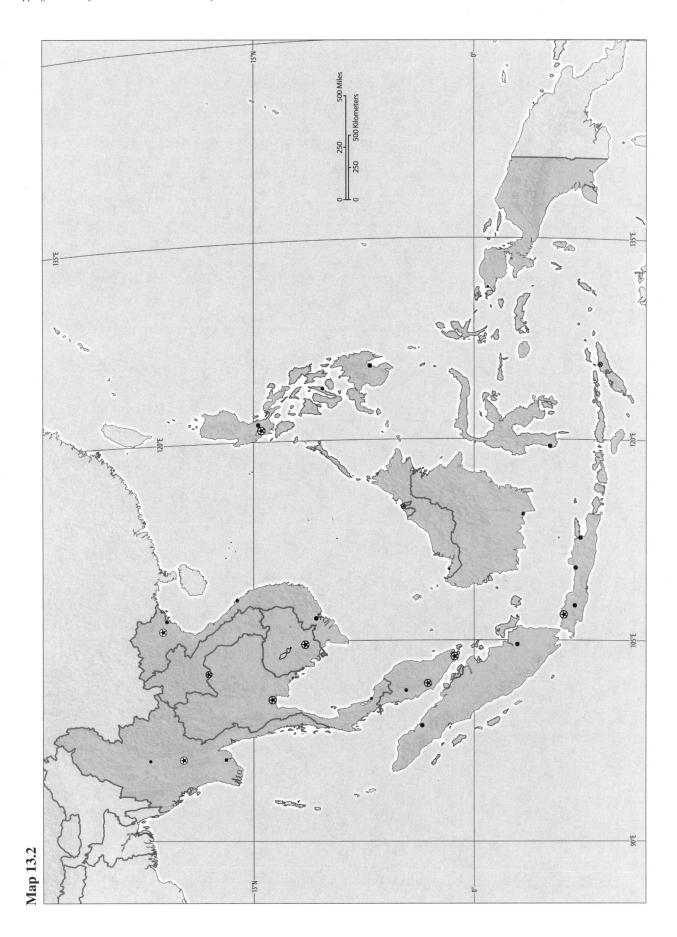

Map 13.2

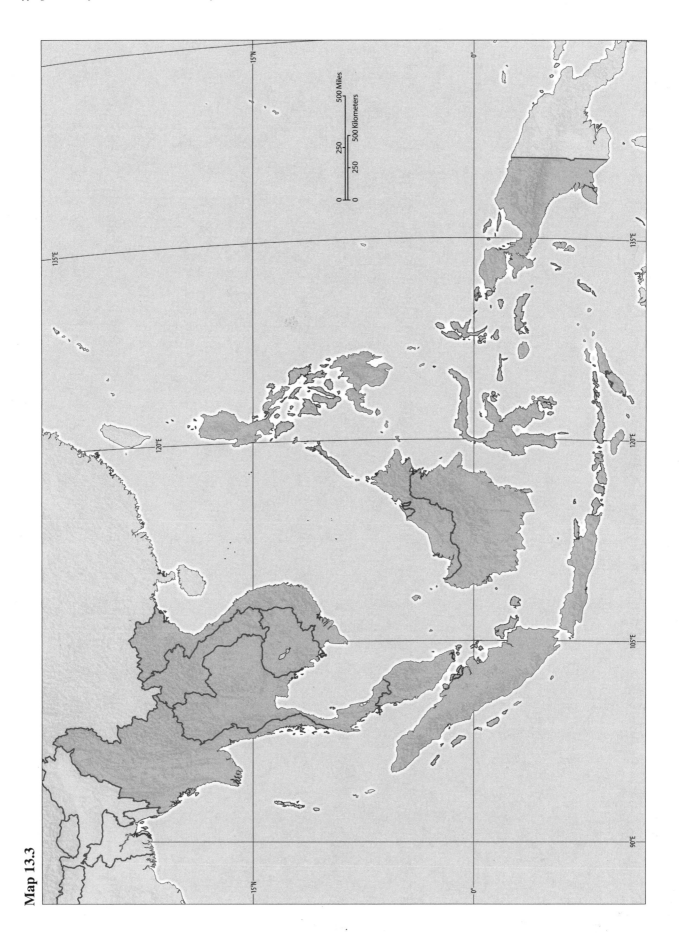

Map 13.3

Exercise One: Southeast Asian Population Settlement and Environmental Issues

Using Figure 13.1 "Southeast Asia" (physical landscape) (p. 427), Figure 13.11 "Population Map of Southeast Asia" (p. 435), Figure 13.3 "Environmental Issues in Southeast Asia" (p. 429), and Mapping Workbook Map 13.4, complete the following exercise.

Using Figure 13.1 "Southeast Asia" (physical landscape) (p. 427) as a reference, use a colored pencil to shade low-elevation regions of both Mainland and Insular Southeast Asia (0–499 feet above sea level). Using a different colored pencil, shade in the higher-elevation regions (500–1,999 feet above sea level). Using a different colored pencil, shade in the higher-elevation regions (2,000–3,999 feet above sea level). Using a different colored pencil from the first two, shade in the highest elevation regions (4,000+ feet above sea level). Mark this shading scheme in the map's legend and label the major mountain ranges. Use a different colored pencil to draw in and label the location of the major Southeast Asian rivers. Once you have done this, answer the following questions.

1. Describe the location of the low-elevation regions in Southeast Asia.

   _____

   _____

2. Describe the location of the higher-elevation regions in Southeast Asia.

   _____

   _____

3. Describe the location of the highest elevation regions in Southeast Asia.

   _____

   _____

Using Figure 13.11 "Population Map of Southeast Asia" (p. 435) as a reference, use a red or black colored pencil to pattern the locations of Southeast Asian population settlement. Mark this patterning scheme in the map's legend. Once you have done this, answer the following questions.

4. Where are the *Mainland* Southeast Asian regions that possess the highest population densities?

   _____

   _____

5. Compare the patterns of elevation, physical features, and population settlement of *Mainland* Southeast Asia that you drew on your map. Do the *Mainland* Southeast Asian physical features seem to play a role in high population densities in this region? If so, what are the physical characteristics of the landscape that appear to promote population settlement?

   _____

   _____

   _____

   _____

6. Where are the *Insular* Southeast Asian regions that possess the highest population densities?

   _____

   _____

7. Compare the patterns of elevation, physical features, and population settlement of *Insular* Southeast Asia that you drew on your map. Do the *Insular* Southeast Asian physical features seem to play a role in high population densities in this region? If so, what are the physical characteristics of the landscape that appear to promote population settlement?

_____

_____

_____

_____

Using Figure 13.3 "Environmental Issues in Southeast Asia" (p. 429) as a reference, use a red or black (or a color that will contrast with the elevation and population shading patterns you established) colored pencil to pattern in the locations of Southeast Asian tropical destruction. Using a different colored pencil, pattern in the locations of the Southeast Asian regions that have experienced the greatest amount of coastal pollution. Mark this patterning scheme in the map's legend. Once you have done this, answer the following questions.

8. Where are the greatest regions of tropical forest destruction in *Mainland* Southeast Asia?

_____

_____

9. Where are the greatest regions of tropical forest destruction in *Insular* Southeast Asia?

_____

_____

10. Where are the greatest regions of coastal pollution in *Mainland* Southeast Asia?

_____

_____

11. Where are the greatest regions of coastal pollution in *Insular* Southeast Asia?

_____

_____

12. Examine the pattern of population settlement in *Mainland* Southeast Asia. Is there a correlation between population settlement and tropical forest destruction? If so, how has the *Mainland* Southeast Asian population affected tropical deforestation in the region?

_____

_____

_____

_____

13. Is there a correlation between population settlement and coastal pollution? If so, how has the *Mainland* Southeast Asian population affected coastal pollution in the region?

_____

_____

_____

_____

14. Examine the pattern of population settlement in *Insular* Southeast Asia. Is there a correlation between population settlement and tropical forest destruction? If so, how has the *Insular* Southeast Asian population affected tropical deforestation in the region?

_____

_____

_____

_____

15. Is there a correlation between population settlement and coastal pollution? If so, how has the *Insular* Southeast Asian population affected coastal pollution in the region?

_____

_____

_____

_____

Exercise Two: Southeast Asian Development Comparison

Using Table 13.2 "Development Indicators" (p. 452) and Mapping Workbook Map 13.5, complete the following exercise.

Using the GNI per Capita, PPP 2010, column in Table 13.2 "Development Indicators" (p. 452) as a reference, use a colored pencil to shade in the Southeast Asian countries possessing an annual per capita GNI of $1,000–$2,500. Using a different colored pencil, shade in the Southeast Asian countries possessing an annual per capita GNI of $2,501–$4,000. Using a different colored pencil, shade in the Southeast Asian countries possessing an annual per capita GNI of $4,001–$5,500. Using a different colored pencil, shade in the Southeast Asian countries possessing an annual per capita GNI of $5,501 and above. Mark your shading scheme in the map's legend.

Using the Human Development Index (HDI), 2011, column from Table 13.2 "Development Indicators" (p. 452) as a reference, use a ruler to draw respective proportional symbols representing the HDI on their respective countries. Have a 1/2 inch (on one side) square represent an HDI of 0.771 or greater, a 1/4 inch (on one side) square represent an HDI of 0.676–0.770, a 1/8 inch (on one side) square represent an HDI of 0.580–0.675, and a 1/16 inch (on one side) square represent an HDI of 0.483–0.579. Use a red or black colored pencil (or a color that will contrast with the shading scheme you created above) to shade in the proportional symbols below.

Study the map you created and answer the following questions.

1. Define GNI.

2. Which Southeast Asian countries possess the highest annual per capita GNI?

3. Which Southeast Asian nations possess the lowest annual per capita GNI?

4. Is there an overall regional pattern of annual per capita GNI for Southeast Asia? If so, what is it?

5. Using GNI as a development indicator, which Southeast Asian region appears wealthier, Mainland or Insular, or are they approximately the same?

6. If one region appears wealthier than the other, what might be the explanation? If the regions are about the same in their levels of development as measured by GNI, what might be the explanation for this similar level of development?

_____

_____

_____

_____

7. Now examine the HDI proportional symbols you placed on your map. Define HDI.

_____

_____

8. Which Southeast Asian nations possess the highest HDI?

_____

_____

9. Which Southeast Asian nations possess the lowest HDI?

_____

_____

10. Is there an overall regional pattern of HDI for Southeast Asia? If so, what is it (e.g., does the region display a sub-region that appears to possess a higher HDI)?

_____

_____

_____

_____

11. Using HDI as a development indicator, which Southeast Asian region appears wealthier, Mainland or Insular, or do they both appear to be about the same?

_____

12. If one region appears wealthier than the other, what might be the explanation? If the regions are about the same in their levels of development as measured by HDI, what might be the explanation for this similar level of development?

_____

_____

_____

_____

13. Compare and contrast the HDI and per capita GNI data you mapped. Is there a correlation (positive or negative) between the HDI and per capita GNI of Southeast Asian nations? If so, what is it?

_____

_____

_____

_____

14. If there was a correlation, what might explain it?

_____

_____

_____

_____

15. If there was no correlation at all, what might explain it?

_____

_____

_____

_____

16. Which of these measurements, HDI or GNI, do you believe to be a better measure of development? Why?

_____

_____

_____

_____

Table

GNI (US$)	Color
1,500–2,500	
2,501–4,000	
4,001–5,500	
5,501 and above	

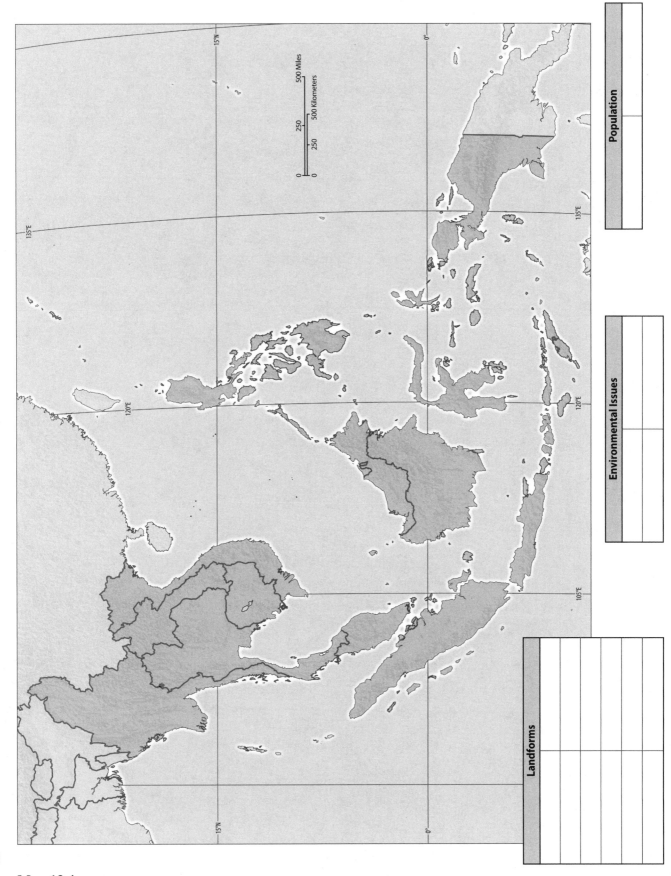

Population

Environmental Issues

Landforms

Map 13.4

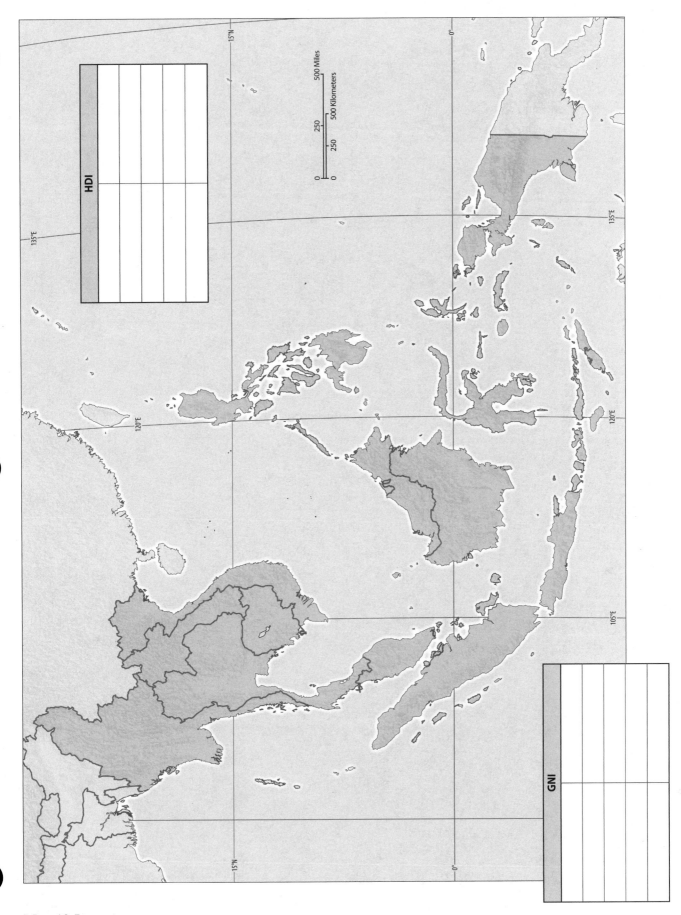

Map 13.5

Chapter Fourteen: Australia and Oceania Mapping Workbook Exercises

Identify the following features on Mapping Workbook Maps 14.1, 14.2, and 14.3

Identify and label the following countries on Mapping Workbook Map 14.1

American Samoa	Marshall Islands	Samoa
Australia	Moorea	Society Islands
Cook Islands	Nauru	Solomon Islands
Federated States of Micronesia	New Caledonia	Tahiti
Fiji	New Zealand	Tokelau
French Polynesia	Niue	Tonga
Guam	Northern Mariana Islands	Tuvalu
Hawaiian Islands	Palau	Vanuatu
Kiribati	Papua New Guinea	Wallis and Futuna
Marquesas Islands	Pitcairn Island	

Identify and label the following cities on Mapping Workbook Map 14.2

Adelaide	Hobart	Papeete
Alice Springs	Honiara	Perth
Auckland	Koror	Port Moresby
Brisbane	Majuro	Port-Vila
Cairns	Melbourne	Suva
Canberra	Noumea	Sydney
Christchurch	Nuku'alofa	Tarawa
Darwin	Pago Pago	Wellington
Funafuti	Palikir	Yaren

Identify and label the following physical features on Mapping Workbook Map 14.3

Arafura Sea	Great Sandy Desert	Pacific Ocean
Bass Strait	Gulf of Carpentaria	Polynesia
Bismarck Sea	Indian Ocean	Solomon Sea
Cook Strait	Kimberly Plateau	South Island
Coral Sea	Line Islands	Spencer Gulf
Darling Ranges	Nullarbor Plain	Tasman Sea
Flinders Range	MacDonnell Range	Tasmania
Great Artesian Basin	Melanesia	Torres Strait
Great Barrier Reef	Micronesia	Tuamotu Archipelago
Great Dividing Range	North Island	

Map 14.1

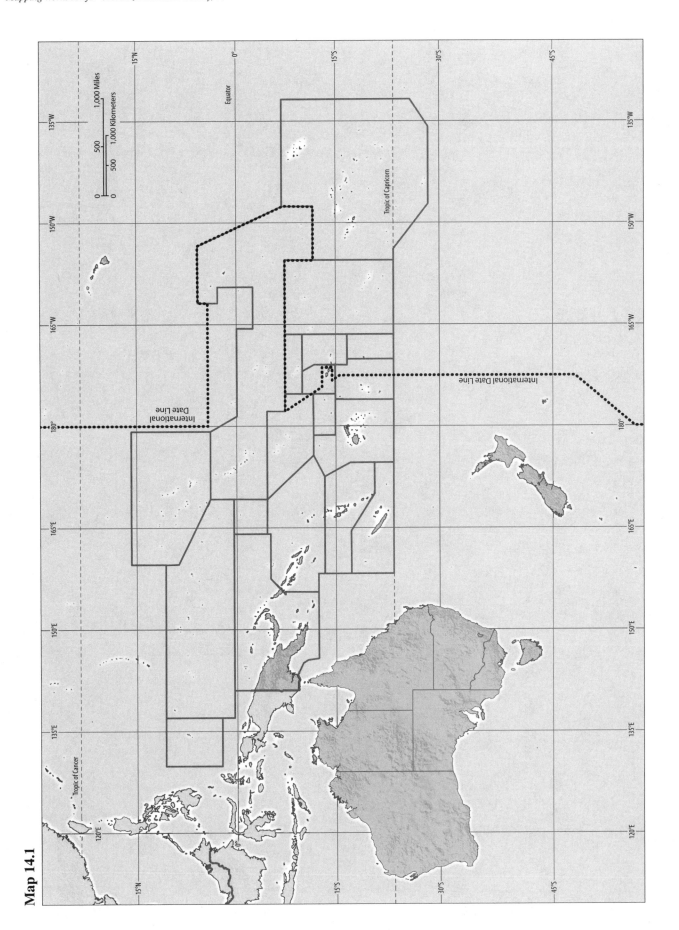

Map 14.2

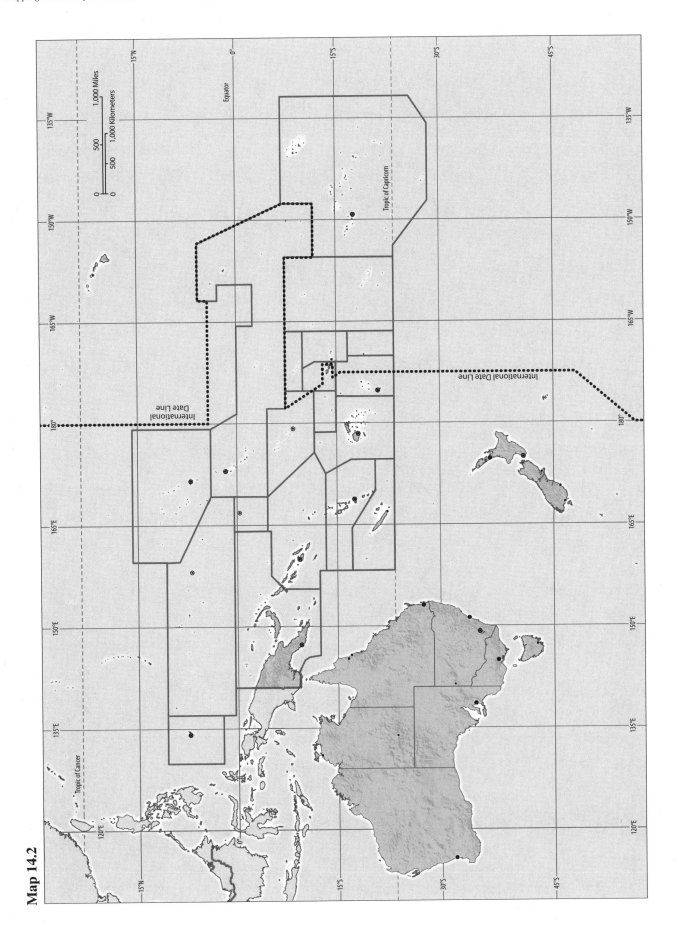

Map 14.3

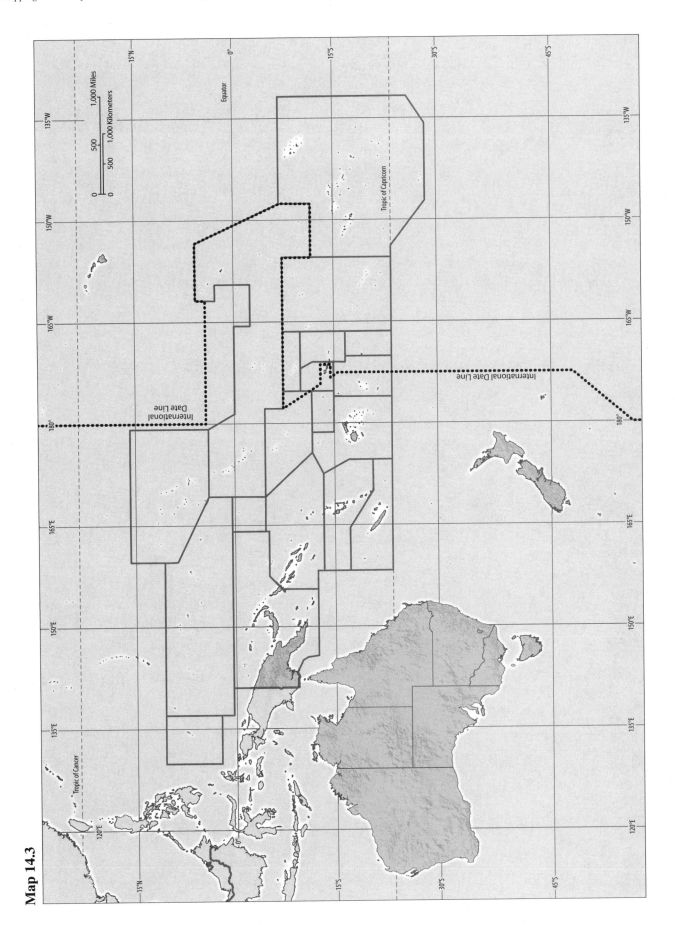

Exercise One: Population and Linguistic Diffusion in Australia and Oceania

Using Figure 14.20 "Peopling the Pacific" (p. 474), Figure 14.32 "Language Map of Australia and Oceania" (p. 482), and Mapping Workbook Map 14.4, complete the following exercise.

Using Figure 14.20 "Peopling the Pacific" (p. 474) as a reference, create an isochronic (line of equal time) map. Begin by using a colored pencil and draw a line around the regions of Australia and Oceania that were settled prior to 2500 BCE. Label this line "2500 BCE." Using a different colored pencil, draw a line indicating the regions of Australia and Oceania that were settled by 1200 BCE. Label this line "1200 BCE." Using a different colored pencil, draw a line indicating the regions of Australia and Oceania that were settled by 200 BCE. Label this line "200 BCE." Using a different colored pencil, draw a line indicating the regions of Australia and Oceania that were settled by 400 CE. Label this line "400 CE." Using a different colored pencil, draw a line indicating the regions of Australia and Oceania that were settled by 800 CE. Label this line "800 CE." Using a different colored pencil, draw arrows indicating the general paths of human migration throughout the region. Mark the line and arrow symbols in the map's legend. Once you have done this, answer the following questions.

1. Describe the extent of settlement in Australia and Oceania by 2500 BCE. Specifically, which groups of islands were inhabited by this period?

    _____

    _____

    _____

    _____

2. Describe the extent of settlement in Australia and Oceania by 1200 BCE. Specifically, which groups of islands were inhabited by this period?

    _____

    _____

    _____

    _____

3. Describe the extent of settlement in Australia and Oceania by 200 BCE. Specifically, which groups of islands were inhabited by this period?

    _____

    _____

    _____

    _____

4. Describe the extent of settlement in Australia and Oceania by 400 CE. Specifically, which groups of islands were inhabited by this period?

    _____

    _____

    _____

    _____

5. Describe the extent of settlement in Australia and Oceania by 800 CE. Specifically, which groups of islands were inhabited by this period?

_____

_____

_____

_____

Using a straight edge, draw a line between point "A" and point "B" on the map. Using the map's scale, determine the distance between these two points. Assuming that point "A" represents the starting location of settlement in this region and point "B" represents the general westward limit of expansion by 800 CE, calculate the average rate of population diffusion in kilometers per century.

6. Between 2500 BCE and 1200 BCE what was the approximate distance that humans had diffused throughout the region?

_____

7. What would be the average rate of diffusion in kilometers per century?

_____

8. Between 1200 BCE and 200 BCE what was the approximate distance that humans had diffused throughout the region?

_____

9. What would be the average rate of diffusion in kilometers per century?

_____

10. Between 200 BCE and 400 CE what was the approximate distance that humans had diffused throughout the region?

_____

11. What would be the average rate of diffusion in kilometers per century?

_____

12. Between 400 CE and 800 CE what was the approximate distance that humans had diffused throughout the region?

_____

13. What would be the average rate of diffusion in kilometers per century?

_____

14. Between 2500 BCE and 800 CE what was the approximate distance that humans had diffused throughout the region?

_____

15. What would be the average rate of diffusion in kilometers per century?

_____

16. During which of these periods did humans most rapidly expand throughout the region?

_____

17. During which of these periods did humans diffuse the most slowly?

_____

Using Figure 14.32 "Language Map of Australia and Oceania" (p. 482) as a reference, use a colored pencil to shade in the regions of Papuan-speaking peoples. Using a different colored pencil, shade in the regions of Austronesian-speaking peoples. Using a different colored pencil, shade in the regions of combined Papuan- and Austronesian-speaking peoples. Using a different colored pencil, shade in the areas of persisting indigenous languages. Mark your shading scheme on the map's legend. Once you have done this, answer the following questions.

18. Where are the majority of the Papuan-speaking people located within Australia and Oceania?

_____

_____

19. Where are the majority of the Austronesian-speaking people located within Australia and Oceania?

_____

_____

20. Where are the majority of the combined Papuan- and Austronesian-speaking people located within Australia and Oceania?

_____

_____

21. Where are the majority of persisting indigenous languages located within Australia and Oceania?

_____

_____

22. Examine the Australian and Oceanic linguistic patterns and compare them with the isochronic lines of diffusion you drew on your map. Based on the mapped linguistic patterns, can you determine the general area of origin and period of expansion of Papuan-speaking peoples throughout the region?

_____

_____

_____

23. Based on the mapped linguistic patterns, can you determine the general area of origin and period of expansion of Austronesian-speaking peoples throughout the region?

_____

_____

_____

24. Based on the mapped linguistic patterns, can you determine the general areas of origin and period when both the Papuan and Austronesian languages became prevalent within the same Oceanic region?

_____

_____

_____

Exercise Two: Comparative Patterns of Oceania Development

Using Table 14.2 "Development Indicators" (p. 492) and Mapping Workbook Map 14.5, complete the following exercise.

Using the GNI per Capita, PPP 2010, column in Table 14.2 "Development Indicators" (p. 492) as a reference, use a colored pencil to shade in the Oceanic countries possessing an annual per capita GNI of $2,000–$3,000. Using a different colored pencil, shade in the Oceanic countries possessing an annual per capita GNI of $3,001–$4,000. Using a different colored pencil, shade in the Oceanic countries possessing an annual per capita GNI of $4,001–$5,000. Using a different colored pencil, shade in the Oceanic countries possessing an annual per capita GNI of $5,001 and greater. Mark your shading scheme in the map's legend.

Using the data from Table 14.2 "Development Indicators" (p. 492) Human Development Index (HDI), 2011, column, take a ruler and draw respective proportional symbols representing the HDI on their respective countries. Have a 1/2 inch (on one side) square represent an HDI of 0.814 and greater, a 1/4 inch (on one side) square represent an HDI of 0.699–0.813, a 1/8 inch (on one side) square represent an HDI of 0.583–0.698, and a 1/16 inch (on one side) square represent an HDI of 0.466–0.582. Use a red or black colored pencil (or another color that will contrast with the shading scheme you have created) to shade in the proportional symbols below.

Study the map you created and answer the following questions.

1. Define GNI.

_____

_____

2. Which Oceanic nations possess the highest annual per capita GNI?

_____

_____

3. Which Oceanic nations possess the lowest annual per capita GNI?

_____

_____

4. Is there an overall regional pattern of annual per capita GNI for Oceania? If so, what is it?

_____

_____

5. Using GNI as a development indicator which, if any, Oceanic region appears wealthier: Micronesia, Melanesia, or Polynesia?

_____

6. If one region appears wealthier than the other, what might be the explanation? If the regions are about the same in their levels of development as measured by GNI, what might be the explanation for this similar level of development?

_____

_____

Now examine the HDI proportional symbols you placed on your map.

7. Define HDI.

_____

8. Which Oceania nations possess the highest HDI?

_____

_____

9. Which Oceania nations possess the lowest HDI?

_____

_____

10. Is there an overall regional pattern of HDI for Oceania? If so, what is it (e.g., does the region display a sub-region that appears to possess a higher HDI)?

_____

_____

11. Using HDI as a development indicator, which Oceanic region appears wealthier: Micronesia, Melanesia, or Polynesia?

_____

12. If one region appears wealthier than the other, what might be the explanation? If the regions are about the same in their levels of development as measured by HDI, what might be the explanation for this similar level of development?

_____

_____

13. Compare and contrast the HDI and per capita GNI data you mapped. Is there a correlation (positive or negative) between the HDI and per capita GNI of Oceanic nations? If so, what is it?

_____

_____

14. If there was a correlation, what might explain it?

_____

_____

15. If there was no correlation at all, what might explain it?

_____

_____

16. Which of these measurements, HDI or GNI, do you believe to be a better measure of development? Why?

_____

_____

Table

GNI (US$)	Color
2,000–3,000	
3,001–4,000	
4,001–5,000	
5,001 and greater	

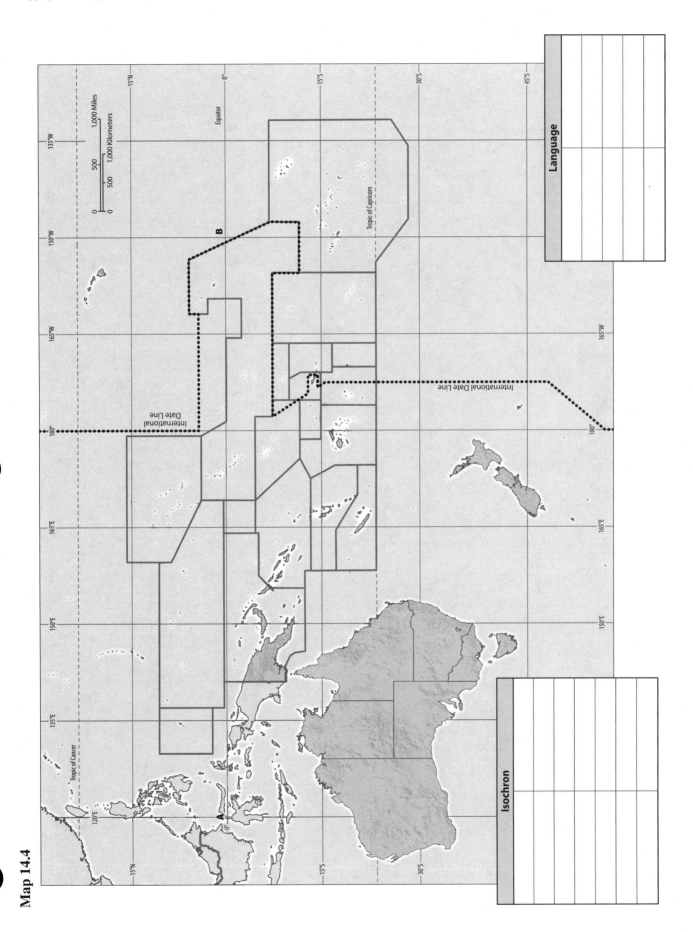

Map 14.4

Language

Isochron

Map 14.5

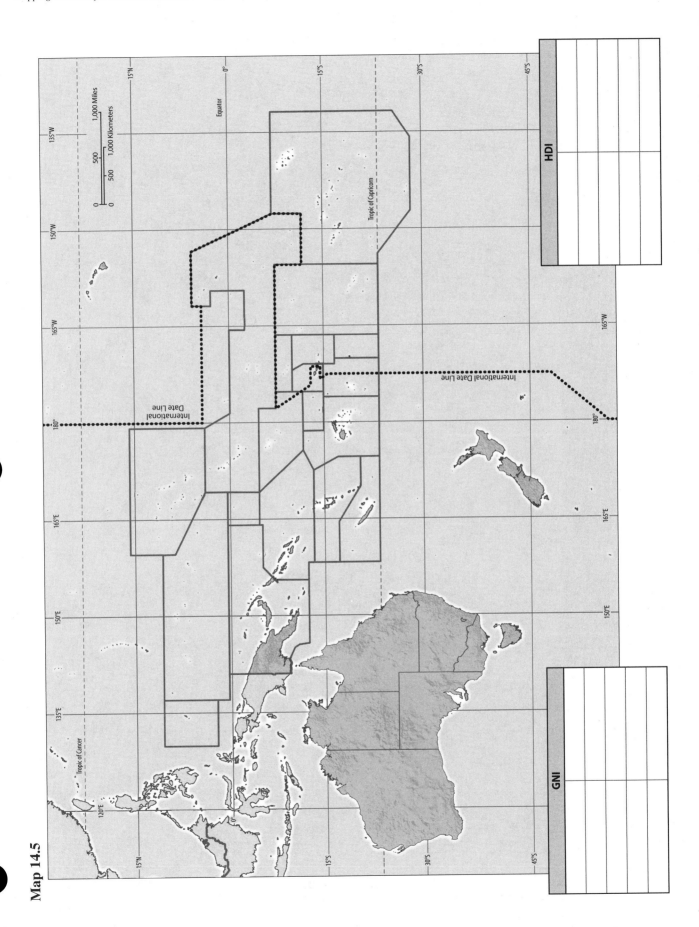